Comet and Asteroid Impact Hazards on a Populated Earth

COMPUTER MODELING

Comet and Asteroid Impact Hazards on a Populated Earth

COMPUTER MODELING

John S. Lewis
*Lunar and Planetary Laboratory
and
Space Engineering Research Center
University of Arizona
Tucson, Arizona*

ACADEMIC PRESS

San Diego   London   Boston   New York   Sydney   Tokyo   Toronto

Cover image © 1999 By David Egge.

This book is printed on acid-free paper.

Copyright © 2000 by ACADEMIC PRESS

All Rights Reserved.
No part of this publication may be reproduced or transmitted in any form or by any means, electronic or mechanical, including photocopy, recording, or any information storage and retrieval system, without permission in writing from the publisher.

Requests for permission to make copies of any part of the work should be mailed to: Permissions Department, Harcourt Brace & Company, 6277 Sea Harbor Drive, Orlando, Florida, 32887-6777.

Academic Press
a division of Harcourt Brace & Company
525 B Street, Suite 1900, San Diego, California 92101-4495, USA
http://www.apnet.com

Academic Press
24-28 Oval Road, London NW1 7DX, UK
http://www.hbuk.co.uk/ap/

Library of Congress Catalog Card Number: 99-62311
International Standard Book Number: 0-12-446760-1

PRINTED IN THE UNITED STATES OF AMERICA
99 00 01 02 03 04 MM 9 8 7 6 5 4 3 2 1

*To the small cadre of dedicated near-earth
asteroid searchers who constitute this planet's
first line of defense against global disaster,
a group that is, in David Morrison's phrase,
"fewer in number than the staff of a small McDonald's"*

*especially including
Tom Gehrels and Jim Scotti of Spacewatch
Glo Helin
Gene and Carolyn Shoemaker
Rob McNaught
Duncan Steel
and the team, as yet unknown, that will find
the Big One.*

Gene, we miss you.

Contents

Preface xi

1
Introduction

Historical Records 1

Quantitative Treatments of Entry and Impact Phenomena 11

Risk Assessment 12

2
The Impact Flux

Mass and Energy Distributions 27

Orbits and Entry Velocities 34

3
The Impactor

Impactor Classes 39

Chemical Properties of Impactors 46

Physical Properties of Impactors 50

4
The Impact Process

5
The Target

Population Density and Distribution 75

Hazards to Populated Areas 77

Hazards in and on the Ocean 77

Property Damage and Destruction 79
Particularly Hazardous Targets 80

6
Method of Calculation

Program Structure 83
The Typical Run 87

7
Results

Results of 100 × 1-Year Runs 89
Results of 1000 × 1-Year Runs 102
Results of Longer Runs 116

8
Implications for Hazard Assessment and Abatement

Assessment of the Impact Hazard 123
Limitations on Detection Strategies 125
Economics of Search and Tracking 130
Hazard Abatement 134

9
Areas Requiring Further Study

10
Conclusions

Appendix A
Program HAZARDS Version 5.5 Owner's Manual   147

Appendix B
Program HAZARDS Version 5.5 Program Listing   155

Appendix C
Program HAZARDS Version 5.5 Sample Output   169

References   187
Index   195

Preface

Awareness of the possibility of large impact events on Earth, although long present among a handful of the most imaginative thinkers, has come of age in this century as a result of studies of Arizona's Meteor Crater and the Tunguska fireball of June 30, 1908, in Siberia, spacecraft observations of cratering on Earth and other rocky bodies, and astronomical surveys of the near-Earth asteroid and comet populations. Appreciation of the effects of large impacts has developed in response to these studies and to the unclassified literature on the effects of large nuclear weapons.

Several earlier volumes have been devoted to various aspects of the impact hazard. The most complete treatment is in the book *Hazards Due to Comet and Asteroids*, edited by Tom Gehrels for the University of Arizona Press Space Science Series (1993), in which issues related to the discovery, tracking, and prediction of near-earth objects (NEOs), their effects on Earth, and possible schemes for interdiction and threat abatement are all reviewed in depth. Other conference volumes within the scope of the present discussion include a detailed report on the 1991 NASA Near-Earth Object Detection Workshop (Morrison, 1992) and the Near-Earth Object Interception Workshop (Canavan *et al.*, 1992). The 1993 Erice workshop, Planetary Emergencies: The Collision of an Asteroid or Comet with the Earth, gave rise to a book project of the same name, which, after collection, editing, and revision of its component chapters, was abandoned unpublished in 1997 by its publisher, Springer-Verlag. The 1995 Planetary Defense Workshop at Lawrence Livermore National Laboratory also produced a collection of reports on related subjects.

In general, hazard modeling has been restricted to very detailed physicochemical treatments of individual events and broad generalizations based on presumably "typical" events. Often only one or two impact velocities and one or two atmospheric entry angles have been considered for each of a small number of different impactor masses. The projectile's properties are often acknowledged by considering two compositional (strength) classes, asteroidal and cometary. Sometimes high-strength materials (iron meteorites) are explicitly included as a part of the impacting population. The target, Earth, has usually been represented by a single globally averaged population density or by a bimodal treatment in which the (unpopulated) oceans are contrasted to the (uniformly populated) continents.

Such treatments give a reasonable estimate of long-term average fatality rates. There are, however, two important reasons why one should not be content with such a level of simulation.

First, it has become abundantly clear that most fatalities on the century or millennium time scale are caused by rare outliers from the general population of impactors. Individual bodies with exceptionally high strengths or exceptionally low entry velocities contribute to fatalities at a rate grossly disproportionate to their numbers. Second, the question of what to expect in the next century, millennium, or 10 millennia is really composed of two distinct parts, only one of which, the mean fatality rate, is answered by time-averaged and spatially averaged treatments. The other crucial issue is the variability of the impact hazard. Only when that variability is taken into account can we speak of most probable outcomes, as opposed to mean outcomes. The distribution of fatalities in impact events is, as with many other types of hazards, a "catastrophic distribution," meaning that the largest single event usually accounts for the majority of the lethality in any time interval. The best way to address both mean fatality rates and the statistical distribution of outcomes is by means of Monte Carlo simulations that employ detailed, realistic mathematical models of the bombarding population, the target human population, and the effects of the impactors.

The purpose of this monograph is to integrate the astronomical, chemical, and physical studies on the bombarding flux and the phenomena of entry and to apply this understanding to the bombardment of a living, densely populated Earth. The time scale of interest to human civilizations ranges from a single human lifetime (order 10^2 years) to the duration of human civilization on Earth (order 10^4 years). On such time scales, the most intensively studied impact phenomenon, impact cratering, is of limited importance, due to the rarity and large mean time between events for crater-forming impacts. Almost all events causing property damage and lethality are due to bodies less than 100 meters in diameter, almost all of which, except for the very largest and strongest, are fated to explode in the atmosphere. Of those that survive the rigors of atmospheric entry all the way to the planetary surface, over 72% must fall into the oceans. Since explosions greater than 1 gigaton TNT are rare on this short of a time scale, we are forced to conclude that the complex behavior of smaller bodies is closely relevant to the threat actually experienced by contemporary civilization.

In this book, the effects of comet and asteroid bombardment on a densely populated Earth (5 billion people) are studied by means of Monte Carlo modeling of the impact flux and of all known hazardous physicochemical correlates of impact. This simulation draws on many discoveries of the solar system exploration program to assess the importance of the impact hazard to Earth. This is the first impact hazard assessment that examines the effects of realistic covariation of impactor mass, strength, composition, orbit, frequency, and entry geometry. Earth is treated in terms of a statistical model that includes not just a correct mean global population density, but, for the purpose of assessing the expected variability of the fatality and damage rate from century to century, a population density distribution model that takes into account modern urbanization and the systematic migration of global population toward seacoasts.

As we shall see, on the century time scale, firestorm ignition and direct blast damage by rare, strong, deeply penetrating bodies are the most common threats to human life, with average fatality rates of about 250 people per year. The most probable outcome for a century is an average fatality rate of about 20 people per year. On average, and neglecting small-meteorite falls that usually affect only a single person, only one or two fatal events per century are expected at modern population densities. On a 1000-year scale, the most severe single event, which is usually a 10- to 100-megaton Tunguska-type airburst, accounts for most of the total fatalities. On longer time scales, regional impact-triggered tsunamis become the most dangerous events. Even larger impacts, above the threshold size for onset of global effects (probably a few gigatons yield), involve bodies that are large enough to be readily discovered and few enough to be affordably tracked and monitored, and even, if necessary, interdicted. The exact impactor threshold size for global effects remains poorly determined.

These findings have implications for our strategies of search, characterization, prediction, and interdiction. They also help focus attention on certain areas, such as stratospheric aerosol injection and tsunami generation and propagation, that require further study. Perhaps most interesting is the implication that the large majority of lethal events (not of the number of fatalities) are caused by bodies that are so small, so faint, and so numerous that the cost of the effort required to find, track, predict, and intercept them exceeds the cost of the damage incurred by ignoring them. The previous general consensus that only the largest impacting bodies, especially those capable of having global deleterious effects, need to be found and tracked is generally upheld, albeit with some caveats regarding regional (neither global nor local) events. Regional events are found to have a higher relative importance in these simulations than in several previous assessments.

1
Introduction

Historical Records

Astronomy books and papers far too numerous to cite offer the assurance that "no one has ever been killed by a meteorite." But the qualitative risk of comet and asteroid impacts has been implicit in human records since the earliest times. Many ancient sources from many cultures treat comets as literal, physical harbingers of doom. Such phenomena as the burning of cities and the overthrow of buildings and walls by aerial events are mentioned many times in Latin, Greek, Hebrew, and Chinese records, but there is no evidence of physical understanding of the nature of the bombarding objects or their effects until quite recently. If these reports are mere baseless superstition, from where did the idea come? What do the ancient sources, casually dismissed by modern American authors who read or trust only reports written in English, actually have to say?

There is indeed a language problem in understanding the ancient reports, but it is largely a matter of the lack of an appropriate technical vocabulary in the older writings. Pliny the Elder, for example, went to some lengths in his *Historia Naturalis* to introduce a terminology for aerial phenomena, but in the absence of quantitative, scientific observations of the same events, the meaning and application of some of his terms remain obscure. Because the cause and true nature of observed entry and impact phenomena were not known to those who observed them, the language used to describe impact events in the older literature must necessarily be "unscientific." In certain locations and periods, especially in medieval Europe, all unusual heavenly events were interpreted as signs sent by God. Therefore, the surviving accounts are strongly biased toward explaining the moral purpose of these events, not their physical nature. Such fundamental information as exact date and time, exact location, place of appearance of the phenomenon in the sky, its duration and physical extent, luminosity, precise nature of the damage done, and the like were generally regarded as unimportant, and therefore rarely recorded for posterity. In China, with its tradition of systematic astronomy organized and recorded by court astrologers, more care was taken with dates and times, and interpretations (which were the product desired by the emperors who paid the salaries of the court) were often perfunctory. Even in 20th century newspapers, bolide explosions may be described (and indexed) as "mysterious explosions," aerial blasts, aerolites, aeroliths, bolides, earthquakes, fireballs, meteorites, meteors, shocks, thunder, and so on. Conversely, the word *meteor* formerly connoted almost anything unusual seen in, or seen to fall from, the sky, including phenomena as diverse as auroral displays and rain.

3

Historical Records

Reports of meteorite falls, often with consequent damage, extend back to the fall of a "thunderstone" in Crete in 1478 B.C., described by Malchus in the *Chronicle of Paros*. The earliest Biblical source is the account of a lethal fall of stones in about 1420 B.C. in Joshua 10:11. The word for stones in this account (*abhagim*) means "stones" or "ores," although some translators render the word as "hailstones" in the passage in Joshua. This emendation may reflect the prevailing academic sentiment regarding the (non)existence of meteorites at the time of translation of the King James version of the Bible.

Other ancient reports in the West are found in the writings of Pausanius, Plutarch, Livy, Pindar, Valerius Maximus, Caesar, and many others. The report of a great fall of black dust at Constantinople in 472 B.C., perhaps the result of a high-altitude airburst, is documented by Procopius, Ammianus Marcellinus, Theophanes, and others.

Colonel S. P. Worden has called to my attention the following passage in *The History of the Franks*, written by Bishop Gregory of Tours: "580 AD In Louraine, one morning before the dawning of the day, a great light was seen crossing the heavens, falling toward the east. A sound like that of a tree crashing down was heard over all the countryside, but it could surely not have been any tree, since it was heard more than fifty miles away. . . .the city of Bordeaux was badly shaken by an earthquake . . . a supernatural fire burned down villages about Bordeaux. It took hold so rapidly that houses and even threshing-floors with all their grain were burned to ashes. Since there was absolutely no other visible cause of the fire, it must have happened by divine will. The city of Orleans also burned with so great a fire that even the rich lost almost everything."

Astronomers who have sought documentary evidence of ancient astronomical phenomena (eclipses, comets, fireballs, etc.) have found that East Asian records are far superior to

European records for many centuries. Kevin Yau has searched Chinese records and found many reports of deaths and injuries (Yau *et al.*, 1994). The Chinese records of lethal impact events (Yau *et al.*, 1994) include the death of 10 victims from a meteorite fall in 616 A.D., an "iron rain" in the O-chia district in the 14th century that killed people and animals, several soldiers injured by the fall of a "large star" in Ho-t'ao in 1369, and many others. The most startling is a report of an event in early 1490 in Ch'ing-yang, Shansi, in which many people were killed when stones "fell like rain." Of the three known surviving reports of this event, one says that "over 10,000 people" were killed, and one says that "several tens of thousands" were killed.

On 14 September 1511, a meteorite fall in Cremona, Lombardy, Italy, reportedly killed a monk, several birds, and a sheep. In the 17th century we find reports of a monk in Milano, Italy, who was struck by a meteorite that severed his femoral artery, causing him to bleed to death, and of two sailors killed on shipboard by a meteorite fall in the Indian Ocean.

In addition to these shipboard fatalities, there have been several striking accounts of near disasters involving impacts very close to ships. Near midnight of 24 February 1885, at a latitude of 37°N and a longitude of 170°15′E in the North Pacific, the crew of the barque *Innerwich*, en route from Japan to Vancouver, saw the sky turn fiery red: "A large mass of fire appeared over the vessel, completely blinding the spectators; and, as it fell into the sea some 50 yards to leeward, it caused a hissing sound, which was heard above the blast, and made the vessel quiver from stem to stem. Hardly had this disappeared, when a lowering mass of white foam was seen rapidly approaching the vessel. The noise from the advancing volume of water is described as deafening. The barque was struck flat aback; but, before there was time to touch a brace, the sails had filled again, and the roaring white sea had passed ahead."

A strikingly similar event occurred only 2 years later on the opposite side of the world. Captain C. D. Swart of the Dutch barque *J.P.A.* reported in the *American Journal of Meteorology* 4 (1887) that, when sailing at 37°39′N and 57°00′W, at about 5 P.M. on 19 March 1887, during a severe storm in which it was "as dark as night above," two brilliant fireballs appeared as in a sea of fire. One bolide "fell into the water very close alongside the vessel with a roar, and caused the sea to make tremendous breakers which swept over the vessel. A suffocating atmosphere, and perspiration ran down every person's face on board and caused everyone to gasp for fresh air. Immediately after this, solid lumps of ice fell on deck, and everything on deck and in the rigging became iced, notwithstanding that the thermometer registered 19° Centigrade."

On 20 August 1907, the steamship *Cambrian* arrived in Boston from England with an equally extraordinary tale to tell. When the ship was several hundred miles south of Cape Race, Newfoundland, steaming along under a clear sky, a brilliant fireball appeared near the northeastern horizon and "rushed across the sky like a rocket. The next moment it passed over the topmast of the liner with a tremendous roar and plowed up the sea about fifty yards from the boat. The upheaval of the water was terrific, but the ship was not damaged." The report of this event was carried in *The New York Times*.

Next, according to the *Times*, on 13 September 1930, a fireball plunged into the sea near Eureka, California, barely missing the tug *Humboldt*, which was towing the Norwegian motorship *Childar* out to sea. It requires little imagination to appreciate that such an event, if it were to strike a ship, should easily cause fatalities, or even the loss of the vessel with all hands.

Early scientific surveys of the classical literature and historical reports on impact events include Troili (1766), Chladni (1818), and an anonymous review known to have been written by David Brewster (1819). The latter two reviews were in part inspired by events such as the fall of a 38-cm-diameter stone at Barbotan, near Roquefort, France, on 24 July 1790, which struck a hut and killed a herdsman and a bullock. Also, in July 1810, a "great stone" fell at Shahabad, India. The fireball burned several villages, presumably by radiant heat and not by means of incandescent solid fragments. Several men and women were killed (*Philos. Mag.* **37**, 236).

The famous Tunguska airburst of 1908 was researched by Gallant (1994), who found reports that two of the Tungus reindeer herdsmen injured by that 10- to 15-Mt explosion had died from their injuries.

Several modern sources collect reports of casualties from impacts. Swindel and Jones (1954) documented a meteorite injury of a human being, and Buddhue (1954) listed several reports of people being struck. Buddhue (1954) gives a list of 27 meteorite strikes on buildings. A number of cases are reported in the *Catalogue of Meteorites* (Graham et al., 1985), and Spratt (1991; Spratt and Stephens, 1992) lists 61 examples of reports of damage or injury caused by meteorite falls. Substantial lists of incidents of meteorite strikes on buildings, cars, animals, and people have been collected (Spratt, 1991; Spratt and Stephens, 1992). Spratt cited the fact that there had been no reports of anyone being struck by a meteorite in the previous 34 years. Immediately after his publication, in 1992, the Mbale meteorite fell in Uganda and struck a boy (Jenniskens, 1994). Numerous other reports of injuries, deaths, property damage, and very close calls are collected in the recent book by Lewis (1996). About 150 such examples as given by Lewis (1997) are listed in Table 1.1. Study of

7

Historical Records

this record reveals that the overwhelming majority of the events involving human injury and death appear to be due to the fall of single stone meteorites. However, the overwhelming majority of the reported casualties are due to a single event (the Chinese fall of 1490 A.D.).

In the present century, two large airburst events have been reported. The first, the 30 June 1908 explosion over the Podkammenaya Tunguska river valley in Siberia, is documented in a number of sources, most notably Krinov (1966). The airburst, at an altitude of about 6 km, had an explosive power close to 10 Mt TNT. The blast leveled and ignited some 2000 km^2 of boreal forest, including trees with trunks more than 1 m in diameter. Eyewitness reports attribute two deaths to injuries suffered in this event. Entire herds of reindeer, totaling perhaps 1000 head, were killed within the blast zone.

The second event, also possibly of multi-megaton yield, struck in the Matto Grosso jungle of Brazil on 13 August 1930. Recent descriptions of this event have appeared in *The Observatory* **115**, 250 (1995) and in *The Sciences* **36**, 14 (1996). No field work has yet been done at the probable site of the latter event, and nothing is known regarding possible casualties in this very remote and sparsely populated region. It has recently been reported that a Brazilian physicist has located a surviving eyewitness to that event.

In addition to documented meteorite falls and apparent airbursts of asteroidal bodies, there have been several reports by astronomers of a population of small asteroidal and cometary bodies that have passed extremely close to Earth. On 27 October 1890, according to the London *Times*, observers at Cape Town, South Africa, saw an extraordinary comet with a tail about a half a degree wide and about 90 degrees long. The comet was visible only from 7:45 to 8:32 P.M., during which time it traversed about 100 degrees of arc. Supposing this was a typical small comet, traveling at about 40 km s^{-1} relative to Earth, then its observed angular rate of 2 degrees min^{-1} implies that the comet must have passed within 80,000 km of Earth, about a fifth of the distance of the Moon.

Another report in *Scientific American Supplement* **67**, 362–363 (1909) tells of a small, round, dark spot that was observed to dash across the disk of the Sun on 21 or 22 July 1896 by William Brooks, the director of Smith Observatory in Geneva, New York. From the angular size (about one-thirtieth of that of the Moon) and from the time taken to cross the face of the Sun (3–4 sec), its distance can be estimated as about 10,000 km. Its speed would then be about 26 km s^{-1}, and its diameter must have been about 3 km. Both numbers could be in error by a factor of 2 or 3; nonetheless, the body must have been very close to Earth, well within the Moon's orbit.

The same article also relates that the Scottish astronomer Sir David Gill, on the night of 29 November 1905, reported a brilliant "fireball" that he watched for 5 min before he lost sight of it in a very hazy sky. A typical fireball would be visible for 1 sec—even 5 sec would be unusual. Another observer, identified only as "Mr. Fuller," kept the same body under observation for 2 hr.

Another fast-moving body, cometary in appearance, was reported independently by two experienced observers, Perrine and Glancy, (1916). Around 9:00 in the evening local time, at the Argentine National Observatory at Córdoba, they noticed a diffuse object, with no visible nucleus, with a tail 8–10 degrees long. The body moved across the sky at an angular rate of about 15 degrees hr^{-1}. Glancy calculated the trajectory of the body approximately from the crude measurements they were able to make over a period of 66 min. She found that the body had approached within about 0.0022 AU of Earth, well within the Moon's orbit.

As it happens, yet another small, fast-moving small comet was seen by observers taking a break at the European Southern Observatory in March 1992 (Smette and Hainault, 1992). A comet with a clearly visible nucleus of about the first magnitude (as bright as Vega) was seen to dash through 20 degrees of arc in only 3 min. The comet had a nearly circular coma about 2 degrees in diameter but no obvious tail. Again taking the most probable flyby speed as 40 km s^{-1}, typical of long-period comets, this comet must have flown by at a distance of about 20,000 km. Remembering that the diameter of Earth is about 13,000 km, this is very close indeed. Now, if the reflectivity of the comet nucleus were similar to that of the Moon (which reflects about 7% of the sunlight which strikes it), we can calculate that a first-magnitude body at a distance of 20,000 km would have to be about 170 m in radius. If we use instead the reflectivity of the nucleus of Halley's comet or of a black, carbonaceous asteroid (about 2.3%), then the radius of the solid nucleus would be close to 300 m. A body this size, with the density of a mixture of ice and rock (about 2 g cm^{-3}), traveling at 40 km s^{-1}, would have a kinetic energy content as high as 15,000 Mt TNT.

In the 1990s, several very close passes of asteroids were observed by the Spacewatch asteroid-search telescope. In May 1996 a newly discovered near-Earth asteroid (NEA), 1996 JA1, was observed to pass within the Moon's distance. This body was about 300 m in diameter, and traveling at 26 km s^{-1}, giving it the kinetic energy of a 2500-Mt explosive. The 1.6-km-diameter Earth-crossing NEA 1997 XF11, discovered by Spacewatch in November 1997, reminds us that numerous Earth-threatening bodies large enough to have severe global effects remain undiscovered. The fraction of Earth-crossing bodies discovered and cataloged to date has been estimated as 10–15% of their actual number. If a major impact were to occur next year, the impacting body would almost certainly be one of the undiscovered majority. Our warning time would be a few seconds, commencing with the body becoming luminous and readily visible at an altitude of 100–140 km.

The ultimate in documented near misses has often been cited as the 1972 fireball that traversed the atmosphere almost horizontally from Utah to Alberta, finally skipping out into a heliocentric orbit that, of course, will intersect Earth again. This event was, however, upstaged on October 8, 1996. Another brilliant fireball was seen over northern New Mexico, traveling almost horizontally through the upper atmosphere in a northeasterly direction. The fireball partially broke up, spraying incandescent fragments, but the main mass survived passage through the upper atmosphere and was last seen fading away as it skipped out. Just 100 min later the same body, after completing one orbit around Earth, again entered the atmosphere off the coast of California, crossed the coastline, and burned up. This is the first documented case of a piece of asteroidal debris being captured into orbit around Earth.

The bombarding flux is real and relatively well understood. The assertion that "nobody has ever been killed by a meteorite" is, to my mind, indefensible. Such a statement requires absurdly high standards of proof, akin to dismissing eyewitness reports of fatal automobile accidents unless the witnesses include both a doctor and an automotive engineer. A more accurate summary of the evidence would be that "nobody has ever been killed by a meteorite in the presence of a physician and a meteoriticist."

As an antidote to this simplistic attitude, I append in Table 1.1 a list of more than 150 reports of damage, injury, death, and near misses collected from Latin, Greek, Chinese, French, German, English, Italian, Japanese, and other sources.

11

Quantitative Treatments of Entry and Impact Phenomena

The first clear understanding that the kinetic energy of an incoming object was responsible for its luminosity, ablation, and fragmentation was expressed by Joule (1848). Writing just 8 years after he first proposed what we now call the First Law of Thermodynamics, Joule described the phenomena associated with the flight and burnup (evaporation) of a meteor in terms of conservation of energy, pointing out that an entering meteor carries a kinetic energy far larger than that needed to crush, melt, and vaporize itself. He pictured the excess energy as being effective in violently heating a mass of atmosphere much larger than that of the meteor. At Joule's time there was still considerable disagreement regarding the nature of meteors. Some, such as Edmund Halley, had argued for an origin of meteors as small solid bodies in heliocentric orbits. Others interpreted them as flammable vapors baked out of the Earth by the Sun, ignited at high altitude. Yet others saw them as electrical phenomena, or as grains of solid ash ejected from volcanoes on the Moon. Joule endorsed Halley's ideas as being better in accord with "modern" physics, such as the concept of vis viva (kinetic energy), and, in accordance with Occam's razor, as offering a more complete explanation of the observed phenomena with fewer untestable assertions or arbitrary inventions.

This first view of entry physics, pioneered by Joule, was brought to a pinnacle by the ubiquitous Ernst Öpik in his treatise, *Physics of Meteor Flight through the Atmosphere*, published 110 years after Joule's great insight (Öpik, 1958a). Öpik, writing at the dawn of the space age, and before the declassification of data on nuclear weapons testing, to a great extent provided the common language and conceptual framework for treating entry phenomena.

In the same time period, the larger significance of impacts in the geological and biological history of Earth was suggested several times, quite independently of one another, by two paleontologists (de Laubenfels, 1956; McLaren, 1970), an astrophysicist (Öpik, 1958b), and a chemist (Urey, 1973). Yet this idea continues to generate determined resistance among some paleontologists, as exemplified by two memorable (and anonymous) polemical editorials in *The New York Times* (17 February 1981 and 2 April 1985).

Beginning in the 1960s, new data became available to cast light on these ideas. First, spacecraft reconnaissance of the solar system uncovered a wealth of data on the cratering histories of the Moon and planets, and aerial and space-based observations of Earth unveiled numerous impact craters on Earth, many very ancient and severely weathered. Then astronomical discoveries, especially the development of a good statistical sense of the frequency of comet passages through the inner solar system and the population and orbital characteristics of near-Earth asteroids, encouraged the development of techniques for determining the physicochemical properties of the bombarding population as well as its size (mass) spectrum.

Risk Assessment

Such recent advances in the discovery and characterization of Earth-crossing solar system bodies and in the theoretical understanding of impact phenomena have led to a great number of recent papers on different aspects of the hazard presented to Earth by these bodies. Statistical averages of many of the threatening phenomena have provided a reliable estimate of the long-term average fatality rate and rate of accrual of property damage (see Chapman and Morrison, 1994; Canavan *et al.*, 1994; American Institute of

Aeronautics and Astronautics, 1995). These studies agree in suggesting time-averaged fatality rates of about 3000 people per year, and property damage rates of several tens of millions of dollars per year. In the long (statistically averaged) view, the damage and lethality totals are dominated by a small number of very rare events.

The time-averaged risk is of actuarial interest and is appropriate for use in the event that our sole knowledge of impact phenomena is statistical in nature, for example, derived only from long-term impact flux estimates. But for public policy reasons, the expected events of the next 100 or 1000 years are of principal interest: Equal in importance to the average risk is the intrinsic statistical variability of the impact process. As the new asteroid search programs accumulate data and as our knowledge of the orbits of real asteroids and comets improves, specific predictions of future Earth-threatening encounters can be made. It will then be possible to predict the precise times at which impacts are threatened in the next century, to identify the most dangerous members of the NEA population over that time span, thus to assess the real threat facing us, not some long-term statistical average. But even then our knowledge of the population of potential impactors will remain incomplete. Some small percentage of the threat must be due to long-period comets that strike Earth on their first observable apparition, with little or no advance warning. This component of the impacting flux must be estimated statistically even if we had complete knowledge of the NEA population.

A further complication is that the changing phenomenology of the impact process with impactor size (and hence with impact times scale) suggests a rich complexity of behavior on the century or millennium timescale that has not been adequately explored in the literature to date, with a disproportionate share of the damage from small bodies (i.e., on the century timescale) contributed by infrequent, atypical bodies and trajectories. It is for these reasons that the present Monte Carlo study was undertaken.

TABLE 1.1
Damage, Injuries, Deaths, and Very Close Calls

Date	Place	Source	Event
c. 1420 B.C.	Israel	Joshua 10:11	Lethal meteorite fall
476	I-hsi and Chin-ling, China	Yau et al. (1994)	"Thundering chariots" "like granite" fell to ground; vegetation was scorched
580	France	History of the Franks	Great fireball and blast; Orleans and nearby towns burned
06/25/588	China	Yau et al. (1994)	"Red-colored object" fell with "noise like thunder" into furnace; exploded; burned several houses
01/14/616	China	Yau et al. (1994)	Ten deaths reported in China from meteorite shower; seige towers destroyed
679	Coldingham, England	Anglo-Saxon Chronicle	Monastery destroyed by "fire from heaven"
764	Nara, Japan	Met. 1, 300	Meteorite strikes house
810	Upper Saxony	Vita Caroli Magni	Charlemagne's horse startled by meteor; throws him to ground
1064	Chang-chou, China	Yau et al. (1994)	Daytime fireball, meteorite fall; fences burned
1321–1368	O-chia district, China	Yau et al. (1994)	Iron rain kills people, animals, damages houses

Date	Location	Reference	Description
1369	Ho-t'ao, China	Yau et al. (1994)	"Large star" fell, starts fire, soldiers injured
2-3/1490	Ch'ing-yang, Shansi, China	Yau et al. (1994)	Stones fell like rain; more than 10,000 killed
10-11/1504	China	Yau et al. (1994)	"Large star" fell with "noise like thunder", garden burned
09/14/1511	Cremona, Lombardy, Italy		Monk killed with several birds, a sheep
1620	Punjab, India	Philos. Mag. [1] **16,** 294	Hot iron fell, burned grass; made into dagger, knife, two sabres
1639	China	Yau et al. (1994)	Large stone fell in market; tens killed; tens of houses destroyed
1648	Ship near Malacca		Two sailors reported killed on board ship en route from Japan to Sicily
1654	Milano, Italy		Monk reported killed by meteorite
8-9 1661	China	Yau et al. (1994)	Meteorite smashes through roof; no injuries
11/07/1670	China	Yau et al. (1994)	Meteorite fall, breaks roof beam of house
10/11/1761	Chamblan, France	Mem. Acad. Dijon 1 C. R. Acad. Sci. **7,** 76	House struck and burned by meteorite
07/24/1790	Barbotan and Agen, Gasc., France	Philos. Mag. [1] **16,** 293	Meteorite crushes cottage, kills farmer and some cattle
06/16/1794	Siena, Italy		Child's hat hit; child uninjured

(continues)

TABLE 1.1 (*continued*)

Date	Place	Source	Event
12/19/1798	Benares, India		Building struck
10/30/1801	Suffolk, England	Times 11/3 3d	"Dwelling-houfe of Mr. Woodroffe, miller, near Horringer-mill, Suffolk, was set on fire by a meteor, and entirely confumed, together with a ftable adjoining."
07/04/1803	E. Norton, England		White Bull public house struck, chimney knocked down, grass burned; flight nearly horizontal
12/13/1803	Massing, Czech.		Building struck
07/1810	Shahabad, India	Philos. Mag. [1] **37**, 236	Great stone fell; five villages burned; several killed
11/10/1823	Waseda, Japan	Met. 1, 300	Meteorite strikes house
01/16/1825	Oriang (Malwate), India		Man reported killed, woman injured by meteorite fall
02/27/1827	Mhow, India	Philos. Mag. [4] **25**, 447	Man struck on arm, tree broken
11/13/1835	Belley, Dept de l'Ain, France	Annuaire (1836)	Fireball sets fire to barn
12/11/1836	Macao, Brazil		Several homes damaged, several oxen killed by meteorite

Date	Location	Reference	Event
1841	Chiloe Archipel., Chile		Fire caused by meteorite fall
5-6 1845	Ch'ang-shou, Szechwan, China	*C. R. Acad. Sci.* **12**, 1196	Stone meteorite damages more than 100 tombs
07/14/1847	Braunau, Bohemia	Yau et al. (1994)	A 37-lb iron smashes through roof into room where three children are sleeping; no serious injuries
10/17/1850	Szu-mao, China	Yau et al. (1994)	Meteorite falls through roof of house
12/09/1858	Ausson, France		Building hit
05/01/1860	New Concord, Ohio		Colt struck and killed
08/08/1868	Pillistfer, Estonia		Building struck
01/01/1869	Hessle, Sweden		Man missed by few meters
01/23/1870	Nedagolla, India		Man stunned by meteorite
12/07/1872	Banbury, England	*Nature* **7**, 112	Fireball fells trees, wall
06/30/1874	Chin-kuei Shan, Ming-tung Li, China	Yau et al. (1994)	Thunderstorm; huge stone fell, crushed cottage, killed child
02/16/1876	Judesegeri, India		Water tank struck
01/03/1877	Warrenton, Missouri		Man missed by few meters
01/21/1877	De Cewsville, Ontario		Man missed by few meters
01/14/1879	Newtown, Indiana	*Paducah Daily News*	Leonidas Grover reported killed in bed (probable hoax)

(continues)

TABLE 1.1 (*continued*)

Date	Place	Source	Event
01/31/1879	Dun-le-Poelier, France	C. Flammarion	Farmer reported killed by meteorite
11-12 1879	Huang-hsiang, China	Yau *et al.* (1994)	Rain of stones; many houses damaged; sulfur smell
11/19/1881	Großliebenthal, Russia		Man reported injured by meteorite
03/19/1887	Barque *J.P.A.*, N. Atlantic	*Am. J. Met.* 4 (1887)	Fireball "fell into the water very close alongside"
11/22/1893	Zabrodii, Russia		Building struck
02/10/1896	Madrid, Spain		Explosion; windows smashed, wall felled
03/11/1897	New Martinsville, W. Virginia	*NYT* 1:4	Man knocked out, horse killed; walls pierced
11/04/1906	Diep River, S. Africa		Building struck
09/05/1907	Hsin-p'ai Wei, Weng-li, China	Yau *et al.* (1994)	Stone fell; whole family crushed to death
12/07/1907	Bellefontaine, Ohio	*NYT* 1:4	Meteorite starts fire, destroys house
06/30/1908	Tunguska valley, Siberia		Two reportedly killed, many injured by Tunguska blast
05/29/1909	Shepard, Texas	*NYT* 1:6	Meteor drops through house
04/27/1910	Mexico	*NYT*	Giant meteor bursts, falls in mountains, starts forest fire

Date	Location	Reference	Description
06/16/1911	Kilbourn, Wisconsin		Meteorite struck barn
06/28/1911	Nakhla, Egypt		Dog struck and killed by meteorite
07/19/1912	Holbrook, Arizona	NYT 8/8/32 17:6	Building struck; 14,000 stones fell; man missed by a few meters
01/09/1914	W. France	NYT 1:7	Meteor explosions break windows
11/22/1914	Batavia, New York	NYT 1:8	Meteorites damage farm
01/18/1916	Baxter, Missouri		Building struck
12/03/1917	Strathmore, Scotland		Building struck
06/30/1918	Richardton, N. Dakota		Building struck
07/15/1921	Berkshire Hills, Massachusetts	NYT 15:2	Meteor starts fire in Berkshires
12/21/1921	Beirut, Syria		Building hit
02/02/1922	Baldwyn, Mississippi		Man missed by 3 m
04/24/1922	Barnegat, New Jersey	NYT 1:2	Rocked buildings, shattered windows, clouds of noxious gas
05/30/1922	Nagai, Japan		Person missed by several meters
07/06/1924	Johnstown, Colorado		Man missed by 1 m
04/28/1927	Aba, Japan		Girl struck and injured by dubious meteorite
12/08/1929	Zvezvan, Yugoslavia	NYT III 1:2	Meteor hits bridal party, kills 1

(continues)

TABLE 1.1 (*continued*)

Date	Place	Source	Event
08/13/1930	Brazil	Obs. 115, 250 (1995) The Sciences 36, 14	Brazilian "Tunguska event"; fire and "depopulation"
06/10/1931	Malinta, Ohio	NYT 6/11 3:4	Blast, crater, smell of sulfur, windows broken in farmhouse; four telephone poles snapped, wires down
09/08/1931	Hagerstown, Maryland	NYT 9/9 14:2	Meteor crashes through roof in Hagerstown
08/04/1932	Sao Christovao, Brazil	NYT 6:5	Fall unroofs warehouse
08/10/1932	Archie, Missouri	NYT 8/13 17:6	Homestead struck, person missed by <1 m
02/24/1933	Stratford, Texas	NYT 3/25 17:1	Bright fireball, 4-lb metallic mass; grass burned
08/08/1933	Sioux Co., Nebraska		Man missed by a few meters
02/16/1934	Texas	NYT 2/17 32:3	Pilot swerves to "avoid crash"
02/18/1934	Seville, Spain	Seattle Times 02/23/34	House struck, burned
09/28/1934	California	NYT 1:3	A pilot "escapes" shower
08/11/1935	Briggsdale, Colorado	NYT 21:2	Man narrowly missed by meteorite
03/14/1936	Red Bank, New Jersey	NYT 3/17 23:3	Meteorite through shed roof
04/02/1936	Yurtuk, USSR		Building struck
10/19/1936	Newfoundland	NYT 10/20 27:7	Fisherman's boat set on fire by meteorite

Date	Location	Reference	Description
03/31/1938	Kasamatsu, Japan	*Met.* 1, 300	Meteorite pierces roof of ship
06/16/1938	Pantar, Phillipines		Several buildings struck
06/24/1938	Chicora, Pennsylvania		A cow struck and injured
09/29/1938	Benld, Illinois		Garage and car struck by 4-lb stone
07/10/1941	Black Moshannon Park, Pennsylvania		Person missed by 1 m
04/06/1942	Pollen, Norway		Person missed by 1 m
05/16/1946	Santa Ana, Nuevo Leon	*NYT* 9:4	Meteorite destroys many houses, injures 28
11/30/1946	Colford, Gloucestershire, UK	*NYT* 7:6	Telephones knocked out, boy knocked off bicycle
02/12/1947	Sikhote Alin, nr. Vladivostok	*NYT* 4/29 14:3	Iron meteorites fall; cratering
09/21/1949	Beddgelert, Wales		Building struck
11/20/1949	Kochi, Japan		Hot meteoritic stone enters house through window
05/23/1950	Madhipura, India		Building struck
09/20/1950	Murray, Kentucky	*NYT* 9/21 33:7	Several buildings struck
12/10/1950	St. Louis, Missouri		Car struck

(continues)

TABLE 1.1 (*continued*)

Date	Place	Source	Event
03/03/1953	Pecklesheim, FRG		Person missed by several meters
01/07/1954	Dieppe, France	NYT 1/9 2:6	Meteorite—blinding explosion, smashed windows
11/28/1954	Sylacauga, Alabama	NYT 86:2 and Met. 1, 125	Mrs. Annie Hodges struck by 4-kg meteorite that crashed through roof, destroyed radio
01/17/1955	Kirkland, Washington	Met. 2, 56	Two irons break through amateur astronomer's observatory dome; one sets a fire
02/29/1956	Centerville, S. Dakota		Building hit
10/13/1959	Hamlet, Indiana		Building hit
02/23/1961	Ras Tanura, Saudi Arabia		Loading dock struck
09/06/1961	Bells, Texas	Met. 2, 67	Meteorite strikes roof of house
04/26/1962	Kiel, FRG		Building hit
12/24/1965	Barwell, England		Two buildings and a car struck
07/11/1967	Denver, Colorado		Building struck
04/12/1968	Schenectady, New York	Met. 4, 171	House hit
04/25/1969	Bovedy, N. Ireland		Building hit
08/07/1969	Andreevka, USSR		Building hit

Date	Location	Reference	Description
09/16/1969	Suchy Dul, Czechoslovakia		Building hit
09/28/1969	Murchison, Australia		Building hit
04/08/1971	Wethersfield, Connecticut		House struck by meteorite
08/02/1971	Havero, Finland		Building hit
03/15/1973	San Juan Capistrano, California		Building hit
10/27/1973	Canon City, Colorado		Building hit
08/18/1974	Naragh, Iran		Building hit
01/31/1977	Louisville, Kentucky		Three buildings and a car struck
05/07/1979	Cilimus, Indonesia	Catalogue	Meteorite fell in garden
05/13/1981	Salem, Oregon		Building hit
11/08/1982	Wethersfield, Connecticut	*JRAS Canada* 85, 263 *NYT* 11/10 1:1 and 1/2/83 I 33:5	Pierced roof of house
06/15/1984	Nantong, PRC		Man missed by 7 m
06/30/1984	Aomori, Japan		Building struck
08/22/1984	Tomiya, Japan		Two buildings hit
09/30/1984	Binningup, Australia		Two sunbathers missed by 5 m
12/05/1984	Cuneo, Italy		Strong explosion, blinding flash; windows broken; daytime fireball "bright as Sun"

(continues)

23

TABLE 1.1 (*continued*)

Date	Place	Source	Event
12/10/1984	Claxton, Georgia		Mailbox destroyed by meteorite
01/06/1985	La Criolla, Argentina		Farmhouse roof pierced, door smashed; 9.5-kg stone misses woman by 2 m
07/29/1986	Kokubunji, Japan		Several buildings hit
03/01/1988	Trebbin, GDR		Greenhouse struck by meteorite
05/18/1988	Torino, Italy		Building struck
06/12/1989	Opotiki, New Zealand		Building hit
08/15/1989	Sixiangkou, PRC		Building hit
04/07/1990	Enschede, Netherlands	*New Sci.*, 6/9/90, 37	House hit by believed fragment of Midas
07/02/1990	Masvingo, Zimbabwe		Person missed by 5 m
1991	Tahara, Japan		Meteorite struck deck of car-transport ship; made crater

Date	Location	Source	Description
08/31/1991	Noblesville, Indiana	Sky & Telescope 4/92	Meteorite fall missed two boys by 3.5 m
08/14/1992	Mbale, Uganda	Met. 29, 246	Forty-eight stones fall; roofs damaged, boy struck on head
10/09/1992	Peerskill, New York	S.E.A.N.	Car trunk, floor pierced by meteorite
10/20/1994	Coleman, Michigan	Met. Pl. Sci. 32, 781	Meteorite penetrated roof of house (1997)
1995	Neagari, Japan		Meteorite penetrated car trunk
04/11/1997	Chambrey, France	Arizona Daily Star, 4/12/97	Meteorite penetrated roof of car; set fire
06/13/1998	Portales, New Mexico	Arizona Daily Star	Meteorite penetrated barn roof
07/12/1998	Kitchener, Ontario	Sky & Telescope 2/99	Meteorite falls 1 m from golfer

Key: Catalogue, A. L. Graham, Catalogue of Meteorites (1985); *JRAS, Journal of the Royal Astronomical Society*; *Met., Meteoritics*; *Met. Pl. Sci, Meteoritics and Planetary Science*; *New Sci, New Scientist*; *NYT, The New York Times*; *Obs, The Observatory*; *S.E.A.N., Scientific Events Alert Network* (presently called *Bulletin of the Global Volcanism Network*).

2

The Impact Flux

Mass and Energy Distributions

The starting point in assessing the impact hazard is to understand the nature of the impactor flux that threatens us. For this purpose we draw on several sources of information. First, we have the cratering data for the terrestrial planets. The crater densities on the planets must first be measured and presented as energy-frequency plots. The impact flux can then be calibrated once the age of each cratered surface is known (see, for example, the review by Neukum and Ivanov, 1994). Unfortunately, direct dating of cratered terrains on other solar system bodies is as yet possible only for the Moon. Once these impact flux statistics are in hand, the impact flux must be corrected for the gravitational focusing effects of the target body. This then provides an estimate of the mean bombardment flux in free space near the heliocentric distance of the target.

Second, astronomical observations of Earth-crossing populations of asteroids and comets can be used to provide an independent estimate of the present impact flux. It is not logically necessary for the mean flux and the present flux to be identical; indeed, several authors have proposed that we are at present in an era of enhanced bombardment. Flux statistics have until quite recently been severely limited in their usefulness by the small number and modest sensitivity of photographic asteroid detection techniques, which generally prevented the discovery of bodies with diameters of less than about 100 m. The Spacewatch program, however, has extended the sensitivity limit so that bodies as small as 5–6 m can be and have been discovered. Such extremely small bodies can be seen only while passing very close to Earth, often much closer than the Moon (Scotti et al., 1991). Statistical arguments, taking into account the fractional areal and temporal sky coverage, have provided firm estimates of the impact flux over the size range of 0.05–10.0 km (Wetherill, 1989; Rabinowitz, 1993; Rabinowitz et al., 1993, 1994). These mass flux estimates can be checked against the integrated flux estimates that arise from the study of cosmic dust accretion by Earth (Love and Brownlee, 1993).

Third, recently declassified data from the Defense Support Program (DSP) satellites and related sensors in geosynchronous orbit (Rawcliffe, 1979; Tagliaferri et al., 1994) have provided us with data on the flux of 1- to 5-m bodies, which are both too rare to be reliably reported by properly instrumented sky-watch programs on the ground, such as the European Network and the Prairie Network, and too small to be detected by existing asteroid search programs. A detailed analysis of these data by Nemtchinov *et al.* (1997) finds the present fireball flux to be in accord with the time-averaged flux determined from cratering statistics. Their model includes calculations of the luminous efficiencies of entering bolides for several different size scales, altitudes, and velocities, considering both H-chondritic and metallic materials. The radiative efficiencies are calculated from a radiative/hydrodynamic model that considers some 167 compounds of the elements Fe, O, Mg, Si, C, H, S, Al, Ca, and Na. These efficiencies for meter-sized bodies are generally of order 0.1, and never higher than 0.28; they are generally higher for larger bodies, and somewhat lower for irons than for stones. At lower altitudes the density of the shock front is higher, and the luminous efficiency at a given velocity is usually higher. Iron bolides traveling below escape velocity have quite low luminous efficiencies, especially at smaller sizes (4.3×10^{-5} at 10 km s^{-1} and 50-km altitude for 10-cm iron projectiles), limited by the evaporation rate and excitation state of the vapor. For the purposes of the present Monte Carlo calculations, bodies smaller than 10 m in diameter are rarely important hazards, and all hazardous bodies in this size range are irons. The luminosity of such bodies when deep in the atmosphere is narrowly constrained by luminous efficiencies ranging from 0.11 to 0.18, and the total luminosity of these small irons is so low that radiant wildfire ignition is not a problem. Larger irons (100 m) can safely be assigned luminous efficiencies of about 0.15 when deep in the atmosphere. Stones in the same size range have average luminous efficiencies of about 0.17, an unimportant difference.

Finally, the flux of bodies smaller than about 1 m has been determined by the European and Prairie networks. Such bodies are seen principally as very bright, weak meteors that disintegrate at high altitudes. These bodies are quite diverse in their physical properties, as discussed in Chapter 3, but very low crushing strengths are favored (Ceplecha, 1988, 1993). Recovered meteorites must derive from a small, strong subset of the fireball population.

The orbits of many fireballs have been determined from photographic tracking data (Ceplecha and McCrosky, 1976; Ceplecha et al., 1993). These orbits clearly associate weak, high-velocity fireballs with cometary orbits. The orbits deduced for individual near-Earth asteroids and comets provide very precise determinations of their encounter velocity with Earth. Enough NEAs are known so that we now have good estimates of the statistical distribution of encounter velocities for each major type of entering body. Our confidence in these velocity distributions is now being enhanced by numerical simulations of the orbital evolution of bodies in NEA orbits (Bottke et al., 1994).

The overall flux curve is modeled as shown in Fig. 2.1. The primary variable is taken to be the impact energy yield Y. The slope $S = (d \log N/d \log Y)$ of the flux curve is a function of the size of the time step dt, varying from -0.6 at time steps greater than 10^4 years to $S = -1.2 + 0.15 \log dt$ for time steps of 1 to 10^4 years, $S = -1.2$ for time steps of 10^{-4} to 1 year, $S = -2.6667 - 36667 \log dt$ for time steps of 10^{-5} to 10^{-4} years, and $-.83333$ for time steps of less than 10^{-5} years. Here Y is the total impactor kinetic energy (megatons TNT) at entry. The mass of the impactor is generated from

$$m = 9 \times 10^{12} \, Y/(30 + 0.72528 \log Y)^2 \qquad (2.1)$$

to correct for the systematic dependence of the mean entry velocity on mass (high-velocity bodies are long-period comets, which are relatively more abundant at larger masses).

The differences between the asteroidal and cometary mass distributions are taken into account by subdividing the impactors

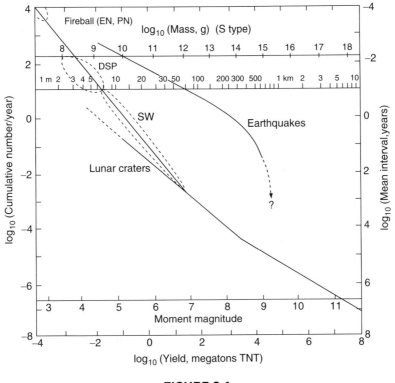

FIGURE 2.1

Size-frequency data on the bombarding flux. Data are drawn from cratering statistics, photographic and electronic asteroid detection searches, comet mass models, and both military space-based (DSP) and scientific ground-based (European Network; Prairie Network) fireball monitoring, as cited in the text. The earthquake magnitude relation treats the seismic energy as equal to the impactor kinetic energy: A realistic efficiency for seismic coupling would be a few percent. Note that earthquakes with moment magnitudes greater than 9.5 or 10.0 can be produced only by impacts.

into cometary and asteroidal components before assigning specific compositions and physical properties to them. Conversion of the impactor yield to the mass and size of the impacting body is, of course, approximate, because the impact velocity of any class of impactor may vary over a wide range. The numbers given at the top of Fig. 2.1 reflect the masses of C- and S-type impactors that strike with typical average velocities for their respective classes. The compositional classes of impacting bodies are treated in detail in Chapter 3.

The probable errors in the mean (time-averaged) flux-versus-yield curve are on the order of a factor of 2 for the largest and smallest bodies. The issue of whether the flux is steady, varying only due to the statistics of small numbers, or genuinely time variable, is unresolved. For our present purposes, it is sufficient to model the flux as steady, and allow the natural statistical variability of the impact rate to proxy for both possible effects. No attempt is made to model clumpiness in the impactor orbital distribution or time correlation of impacts. The existence of orbital "streams" of comet fragments has been suggested (Dorman *et al.*, 1978; Bailey *et al.*, 1993; Steel *et al.*, 1991) and is clearly qualitatively plausible, but the data needed for modeling the time dependence of the impact flux are lacking. Comet breakup has been reviewed by Sekanina (1982), and Schenck and Melosh (1993) have demonstrated the existence of impact crater chains on Callisto due to breakup of comets after close passages by Jupiter. Drummond (1990) has also pointed out several groups of near-Earth asteroids (NEAs) with correlated orbits and, presumably, genetic connections, but these bodies are not numerous enough to bias the statistical distribution of orbital parameters and entry velocities. Rabinowitz *et al.* (1993) suggest an NEA belt.

Figure 2.1 also contains information on the energy release from earthquakes. Earthquake energy release is no longer based on the Richter scale, but rather on the moment magnitude scale, in which each magnitude corresponds to a factor of $10^{1.5}$

in energy. There is a natural limit on the moment magnitude of natural (geologically generated) earthquakes imposed by the maximum strain energy that can be stored on a large fault system. The limit appears to lie somewhat below a moment magnitude of 9. That limit is represented in Fig. 2.1 as the energy release (Y) corresponding to the moment magnitude. Impacts have no such limitation on their total energy content; however, the fraction of the impact energy that is delivered as seismic waves is a small and variable percentage of the total kinetic energy of the impactor. In using Fig. 2.1, caution should be exercised when comparing the earthquake energy curve to the graphed total energy curve for impactors: The seismic effect of a surface impact will almost always be >0.7 to 1.5 magnitudes smaller than the total impactor energy, depending on the densities of projectile and target and on whether the projectile strikes land or water. Thus an impactor with a total energy content equivalent to a moment magnitude of 10.5 may actually deliver no more seismic effect than the very largest natural earthquake. Glasstone and Dolan (1977) cite an efficiency of conversion of explosive yield to seismic shock front energy of less than 5%. Hills and Goda (1993) mention that the coupling between impactor yield and seismic energy is poor. Toon *et al.* (1995) argue, based on the data presented by Schultz and Gault (1975), that the coupling constant is only about 0.01 for impacts in hard rock (basalt), and in the range of 10^{-3} to 10^{-5} for impacts in less competent terrain. Toon *et al.* adopt a figure of 10^{-4} for the conversion of impact energy to seismic surface wave energy, with the understanding that this number is highly uncertain. Nonetheless, even if one takes a coupling constant of 10^{-3}, impact-induced earthquakes larger than a moment magnitude of 9 would occur no more frequently than once every several million years. In the context of the present simulations, covering human to cultural timescales of 10^2 to 10^4 years, the largest seismic disturbances would be those due to Earth's own earthquake activity.

Orbits and Entry Velocities

In the absence of more detailed information, three orbital classes of impactors are recognized in this simulation: stony asteroids, short-period comets, and long-period comets. The definitions of these classes are somewhat different from those usually employed in the literature (asteroids, short-period comets, long-period comets). The differences are that the few high-velocity (retrograde) short-period comets reported by Chyba (1991) are lumped with the long-period comet population in this simulation, and the sole high-velocity NEA (40 km s^{-1}) is treated as a short-period comet. The Earth-impact velocity distributions (km s^{-1}) generated randomly in this simulation are:

$v = 12.4 - 30[1 - \cos(\text{Rnd}^2)]$ (for near-Earth asteroids), (2.2a)
$v = 24 \pm 10\ \text{Rnd}^2$ (for periodic comets), (2.2b)
$v = 46 \pm 27\ \text{Rnd}^2$ (for long-period comets), (2.2c)

where Rnd is a random number between 0 and 1. These synthetic distributions are compared to the observational data for asteroids and periodic comets reported by Chyba (1991) and by Bottke *et al.* (1994), and the impact velocity distribution for long-period comets deduced from data in Marsden and Steel (1994) in Fig. 2.2. In all three parts of Fig. 2.2, the lowest possible entry velocity (Earth's escape velocity) and the highest possible entry velocity (head-on collision with a retrograde parabolic comet) are indicated by vertical dashed lines.

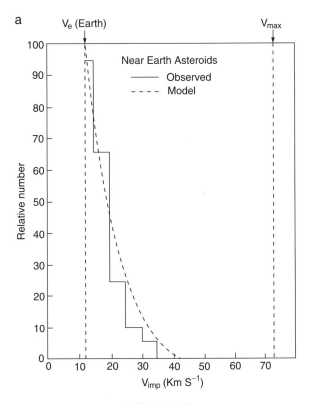

FIGURE 2.2

Velocity distributions of impactors. (a) Comparison of the observed distribution for near-Earth asteroids with the randomly generated velocity distribution in the Monte Carlo model used in Program HAZARDS. (b) Similar comparison of the entry velocity distributions for the periodic (short-period) comets. For parts (a) and (b) the observed distributions are taken from Chyba (1991). (c) Summary of the velocity data for the long-period comets. (Observed distribution from Marsden and Steel, 1994.)

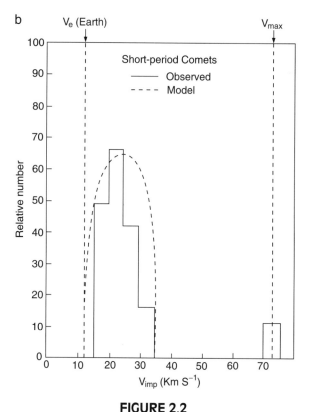

FIGURE 2.2

(continued)

The only noticeable systematic difference between the observational results and the model is the skewed observed velocity distribution for long-period comets. The simulation assigns too few long-period comets to impact velocities between 25 and 45 km s^{-1} and too many in the 50–70 km s^{-1} range. The effect of this difference is that the simulation generates long-period comets with too little energy content, but, because

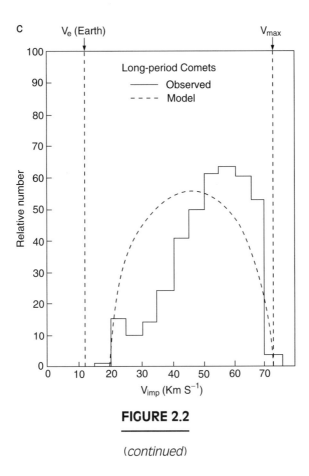

FIGURE 2.2

(continued)

of their lower entry velocities, they penetrate a little deeper into the atmosphere before exploding. Calculations suggest that these two effects offset each other almost perfectly, and contribute no discernible error to the hazard calculations.

3

The Impactor

Impactor Classes

The impactor flux is readily divided into several compositionally and physically distinctive components. The first of these is the near-Earth asteroid (NEA) component, which naturally subdivides into strong (commonly meteorite-producing) bodies from the inner asteroid belt and weak carbonaceous (rarely meteorite-producing) bodies. There are also short-period ("periodic") comets ($P < 100$ years; prograde, modest-inclination, modest-eccentricity orbits) and long-period comets ($P > 10,000$ years; random inclinations and very high eccentricity). The periodic comet flux is discussed by Shoemaker *et al.* (1994) and the long-period comet flux by Marsden and Steel (1994).

For the stronger classes of impactors we draw on our laboratory knowledge of the chemical and physical properties of meteorites. For the weaker bodies, which could not survive entry into Earth's atmosphere, we use properties deduced from studies of fireballs with similar orbital properties (Ceplecha, 1988, 1993; Ceplecha and McCrosky, 1976; Ceplecha et al., 1993).

The NEA population, at present consisting of some 400 known bodies of all sizes from about 5 m to several kilometers in diameter, has been only very imperfectly characterized by photometric and spectroscopic techniques. The historical lack of coupling between asteroid search programs and spectral characterization programs has resulted in many NEAs remaining uncharacterized until one or more apparitions after their date of discovery. Diligent attempts to inform observers promptly of the discovery of new NEAs, using modern electronic communication links, have significantly improved the coordination between discovery and spectral characterization during the past few years; however, the increased sensitivity of Spacewatch has resulted in the discovery of many small asteroids that are too faint for even UBV photometry. The smallest of these bodies are so small that they can only be detected when very close to Earth, which, at typical encounter velocities, may make the observation window too short for adequate follow-up. Nonetheless, some 55 NEAs have been spectrally characterized to some degree in published works (Table 3.1), and unpublished data exist on about another 40.

Of the NEAs that have been studied, some 25% have the very low albedos characteristic of C-type belt asteroids and carbonaceous meteorites. However, to appear in such a list, an asteroid must be both bright enough at visible wavelengths to be discovered and bright enough to be studied photometrically. If NEAs were discovered by IR sky mapping (using a search instrument similar to the IRAS spacecraft, but programmed for asteroid

TABLE 3.1

Frequency of Observed Spectral Types of Near-Earth Asteroids[a]

Type	Number known	Score	Percent	Possible meteorite analogues
S	16	20.0	50.0	Pal.; stony-iron; Ur; CV/CO
SQ	2			
SU	2			
SG	1			
SMU	1			
CSU	1			
QRS	1			
C	6	9.5	23.75	CI and CM
CF	1			
CSU	1			
XC	1			
D	1			(Super-carbonaceous)
F	1			(Flat C-like spectrum)
G	0			(Altered CI/CM)
SG	1			
T	0			(Super-carbonaceous?)
TCG	1			
V	3	3.0	7.5	Basaltic achondrite (Eu,Ho,Di)
M	2	2.5	6.25	Irons; possibly EH and EL
SMU	1			
Q	1	2.5	6.25	H, L, and LL chondrites
QRS	1			
QU	1			
SQ	1			
A	1	1.0	2.5	Pal. or olivine-rich achondrite
E	1	1.0	2.5	Enstatite achondrite (EA)
R	0			Pyroxene-olivine achondrite
QRS	1	0.5	1.25	
U	0			Unknown affinity
SU	1			
SMU	1			
CSU	1			
QU	1			
		40.0	100.0%	

[a] These bodies must be discovered and characterized.

search and follow-up instead of deep-sky mapping) this limitation would not apply; however, NEAs are in fact discovered by visual-wavelength observations. Both of the criteria mentioned discriminate against intrinsically dark asteroids. Attempts to correct the observed NEA statistics for this observational bias suggest that the real proportion of very dark bodies (Bond albedos less than about 0.05) is close to 50%, not 25%.

Also, we should not be too hasty in attributing all dark NEAs to the C-type belt asteroid population. Studies of the provenance of NEAs have suggested that roughly half of the present NEA population is derived from the belt via the wandering of asteroids into Jupiter resonances, with consequent pumping-up of their eccentricities until they cross Earth. The only other major source of NEAs now known is the short-period comets. These periodic comets are in low-inclination prograde orbits, usually very strongly influenced by Jupiter. Long-term orbital and thermal evolution of these bodies produces a residue of bodies whose outer meter or so of surface has been baked free of volatiles by solar heating during numerous perihelion passages (Mendis and Brin, 1977; Brin and Mendis, 1979; Brin, 1980; Fanale and Salvail, 1984, 1987; Horanyi *et al.*, 1984; Houpis *et al.*, 1985; Prialnik and Bar-Nun, 1988). Their surfaces, composed of moderately heated ice-free cometary materials, are dominated by a lag deposit of carbonaceous and mineral dust with very low albedo, high opacity, and low thermal conductivity (Podolak and Herman, 1985). This very dark material must presumably be spectrally similar to C, D, or P asteroids; however, we have such limited knowledge of the spectral properties of solid cometary nucleus material that we simply cannot be sure what to expect. The thermal models of periodic comet evolution, however, agree in predicting the existence of such a porous, dark dust layer, with thickness of tens of centimeters to a few meters, overlying and insulating an ice-rich interior of little-altered com material (McKay *et al.*, 1986; Mekler *et al.*, 1990; Mekler, 1991). We already have one direct demonstration of the presence of a recently deceased short-period comet nucleus, the former comet Wilson-Harrington 1949, masquerading as the asteroid 1979 VA in the NEA population. The relationships between comets and asteroids have been surveyed by Wetherill and ReVelle (1982), Degewij and Tedesco (1982), and Hartmann *et al.* (1987).

An independent method of assessing the population statistics of bodies in Earth-crossing orbits is to examine the taxonomy of meteorite falls. It is not surprising that the statistics of meteorite classes derived from meteorites accidentally discovered on the ground ("finds") differ from those of meteorites that are observed to fall ("falls"): Some rare meteorite classes, such as irons and pallasitic stony-irons, have very low weathering rates because of their domination by natural stainless steel. They are therefore grossly overrepresented among finds compared to falls.

But even the best available statistics of falls (Table 3.2) leave much to be desired as representatives of the incident population at the top of the atmosphere: Several classes of stony meteorites are so fragile that they barely survive entry. Indeed, we frequently observe falls in which enormous numbers of stones, from hundreds to perhaps as many as 100,000, fall together in a compact elliptical "strewn field." Such "meteorite showers" (of course, unrelated to meteor showers) attest to the thorough fragmentation of large incoming bodies by aerodynamic forces. Stagnation point pressures adequate to crush a large incoming body into thousands of kilogram-sized pieces are only slightly less than those needed to crush the body into unobservably (or, more to the point, unrecoverably) small pieces. Thus both the fragility of some stones, especially carbonaceous chondrites, and the ob-

TABLE 3.2

Frequency of Observed Falls of Meteorite Types[a]

Stones	95.7%	Chondrites	87.8%	EH + EL	3.5%
				H	27.0
				L	40.8
				LL	9.0
				CV + CO	3.5
				CM	3.0
				CI	1.0
		Achondrites	7.9%	Eu	2.7%
				Ho	2.4
				Di	1.1
				Ur	0.4
				EA	1.1
				Lunar	0.1
				SNC	0.1
Stony-irons	1.1%			Mes	0.8%
				Pal	0.3
Irons	3.2%			IAB	0.6%
				IIAB	0.4
				IIC	0.1
				IID	0.1
				IIE	0.1
				IIIAB	1.1
				IIICD	0.2
				IIIE	0.1
				IIIF	0.1
				IVA	0.3
				IVB	0.1

[a] These bodies must survive entry and be recovered fresh.

served falls of showers of thoroughly crushed chondritic stones attest to powerful selection effects during atmospheric entry. Correcting the meteorite fall statistics for this effect is qualitatively easy: Weak materials such as carbonaceous chondrites must be much more abundant in the bombarding flux than among the observed falls. But correcting for this effect quantitatively using only information on bodies that reach the ground successfully is very difficult. It is therefore useful to consider the properties and statistics of meter-sized bodies observed by optical fireball tracking networks: The large majority of these bodies are very weak and disintegrate at high altitudes.

The incident population statistics are a function of size. Cometary detritus, in the form of dark, weak, porous carbonaceous "dustballs" in cometary orbits, is abundant at the meter size level. These dustballs may be fragments of the outgassed surface lag deposits of black, friable dust predicted by the models of cometary evolution cited earlier. Because of the fragility of such material, ejection by gentle outgassing of the comet nucleus seems more likely than ejection by violent impacts, since the latter would almost certainly destroy these fragile clumps of dust. There is no reason, either theoretical or observational, to believe that these dustball bodies are important at sizes above a few meters.

On the other hand, true ice-bearing cometary nuclei, including those now indistinguishable from asteroids, are surely present at size scales of a hundred meters or more. It is quite unclear how important such bodies are in the 10- to 100-m size range. There are, as we related in Chapter 1, a few almost anecdotal reports of observations of tiny, faint comets with enormous angular rates, suggestive of a significant population of such "microcomets" (Anonymous, 1890; Perrine, 1916; Glancy, 1916; Smette and Hainault, 1992). There are, however, no useful constraints on the abundance of such bodies or on their sizes. We shall rather arbitrarily let the proportion of such bodies in the range from 10 m up to masses of 4.5×10^{16} g be 25%. Above that mass, to conform with the asteroid statistics cited earlier and the comet data of Hughes (1990), the ratio of the number of asteroids to the number of comets is given by

$$\log(N_{ast}/N_{comet}) = 5.38 - 0.294 \log M \qquad (3.1)$$

up to 2×10^{18} g, beyond which the ratio is taken to be 1.0.

Note that these "microcomets" are in no way related to the small 10- to 100-m ice bodies proposed by Frank and Sogwarth (1993). Frank's proposed bodies partake of a number of improbable traits, such as a complete absence of entry luminosity upon encounter with Earth's atmosphere (i.e., no alkali metals, alkaline earths, or transition metals, no carbon, and no other material with high quantum efficiency for emission of visible light), no detection by photographic and electronic optical search systems (carbon- and mineral-free ice should be highly reflective), no impact signature on the Moon or elsewhere, and fluxes 10^4 to 10^6 times those given for comparable-sized bodies in Fig. 2.1.

Chemical Properties of Impactors

Few chemical properties are of direct relevance to this modeling effort. I distinguish only four such properties as important. First, there is the content and composition of metallic iron. The metallic iron contents of the major classes of impactors range from >99% for irons through about 50% for stony-irons down through 10–30% for most chondritic stones to <1% for carbonaceous chondrites and achondrites, and zero for cometary solids. The metal component is important because it strongly influences the density of the body and provides it with a distinctive signature of rare heavy metals that, on Earth, were very efficiently extracted

Chemical Properties of Impactors

into the core during core formation, and hence are extremely rare on Earth's surface. The type example of this class of elements is iridium. The bulk iridium content of many classes of impactors is summarized in Table 3.3. Further, the iron content enhances

TABLE 3.3

Iridium Abundances in Impactors

Class	Iridium (ppm wt)
Comets	
Long period	0.3[a]
Short period	0.3[a]
Chondrites	
Carbonaceous CI	0.5
Carbonaceous CM	0.6
Enstatite E	0.8?
Ordinary H, L, LL	0.8
Stony-Irons	
Pallasites (0.01–0.1)	0.03
Mesosiderites (1-3)	2.0
Soroti (a unique metal/Fes meteorite)	0.06
Irons (average)	3
4.5% of all irons	0.010–0.020
5.0%	0.020–0.050
2.5%	0.050–0.100
3.5%	0.10–0.20
7%	0.20–0.50
7%	0.50–1.00
17%	1.00–2.00
29%	2.00–5.00
11%	5.00–10.0
10%	10.0–20.0
4%	20.0–50.0
0.5%	50.0–100.0
Achondrites (0.001–0.01)	0.003

[a] Based on a compositional model; not measured.

the luminous efficiency of rock vapor because of the rich line spectrum of atomic iron in the visible spectrum.

The second compositional feature of interest is the sulfur content (Table 3.4). As we shall see, high-altitude disruption of large, weak bodies can inject quite important quantities of sulfur into the upper atmosphere. As I first suggested at the Erice workshop in 1989, this sulfur, efficiently oxidized to sulfate by local atmospheric chemistry, serves as a major source of stratospheric aerosols, which, in extreme cases, may materially increase the overall albedo of Earth and lead to global cooling.

It should be mentioned that the impactor itself is not the only source of sulfur. The Chicxulub impact in Mexico, which occured in a shallow-sea sedimentary terrain dominated by calcite

TABLE 3.4

Sulfur Abundances in Impactors

Class	Sulfur (wt %)
Comets	
Long period	6
Short period	6
Chondrites	
Carbonaceous CI	5.9
Carbonaceous CM	3.4
Enstatite E	4.5
Ordinary H, L, LL	2.0
Stony-Irons	
Pallasites (0.04–2.9)	1.0
Mesosiderites	0.1[a]
Soroti	15.0
Irons	1.0[a]
Achondrites	0.03[a]

[a] Very little data.

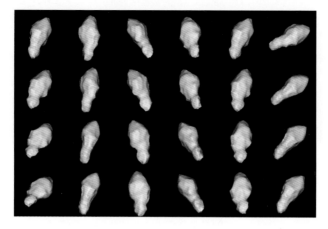

These radar images of the near earth asteroid 4179 Toutatis illustrate the highly nonspherical nature of small solar system bodies. Such bodies would readily fragment upon atmospheric entry. Radar image courtesy of Steven J. Ostro. Reprinted from *Icarus 137*, 130. ©1999 Academic Press.

Earth poisoned and defoliated by nitrogen oxides from the Cretaceous/Tertiary boundary impact event. Brown-red nitrogen dioxide gas oxidizes and hydrates to make nitrogen tetroxide, nitric acid, and nitrous acid, a witches' brew of caustic, toxic, carcinogenic, teratogenic, and mutagenic gases. The gas blocks transmission of all but red light to the ground, curtailing photosynthesis, defoliating trees and shrubs, and killing surface-dwelling animals. Painting by David Egge.

A long-period comet passes close by Jupiter, risking both impact with the giant planet and severe disturbance of its orbit by Jupiter's powerful gravity. Such encounters sometimes (as with Comet Shoemaker-Levy 9) result in temporary capture by Jupiter, tidal disruption of the comet, or transfer to a short-period orbit that takes the comet through the inner solar system every few years. Painting by David Egge.

A typical near-Earth asteroid (NEA), with shattered interior and heavily cratered surface, is visited by an exploratory spacecraft in this painting by David Egge. Some 2000 kilometer-sized asteroids cross or graze Earth's orbit. Of these, about 400 will eventually collide with Earth, threatening mass destruction.

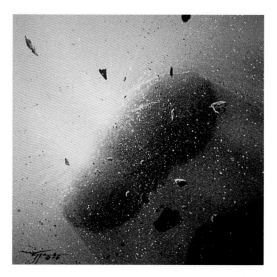

An artist's conception of the surface of the nucleus of Halley's Comet. Streams and jets of gases from evaporating ices levitate and expel dust and larger pieces of surface material. The nucleus may contain large, relatively strong chunks of "permafrost" loosely bound together by the comet's feeble gravitation. After heating by the Sun, the nucleus may be left with a black, fluffy layer of involatile dust on its surface. Painting by David Egge.

A Tunguska-sized impact of a small comet. Depending on its size and the strength of its component pieces, a comet may either disrupt at high altitudes or penetrate to Earth's surface. High-altitude explosions, which are more probable, can ignite fires out to distances of hundreds of kilometers from the explosion point. Painting by David Egge.

A serial view of the fragmentation and ablation of a meteor in Earth's upper atmosphere, with shock fronts and eroded (and rapidly decelerating) fine particles. Fragmentation, ablation, and luminosity are sensitive functions of the size and physico-chemical properties of the entering object. Painting by David Egge.

The impact of the asteroid that ended the Cretaceous Era. A broad conical shock front blasts aside air, ocean, and rock as the intensely hot, dense fireball scours out a crater nearly 200 km in diameter. This crater, on the northern slope of the Yucatan peninsula in Mexico, has been intensively studied in recent years. In 1998 a small fragment of the asteroid itself was recovered from a deep-sea core drilled through the K-T boundary in the Pacific. Vapors and dust from the impact were hurled ballistically over the entire globe within about 45 minutes of the impact. Painting by David Egge.

($CaCO_3$) and anhydrite ($CaSO_4$), must have released vast quantities of sulfur dioxide into the atmosphere via shock heating and devolatilization of the target rocks (Sigurdsson *et al.*, 1992; Chen *et al.*, 1994). Brett (1992) has estimated that some 4×10^{17} g of sulfur dioxide was emitted, whereas Pope *et al.* (1994) place the figure from as low as 8×10^{16} g to as high as 1.4×10^{18} g. Sulfur is also injected by ocean impacts, but, owing to the low concentration of sulfates in sea water, the contribution of marine sulfates is only important for exceptionally sulfur-poor impactors such as achondrites. Kring *et al.* (1996) have discussed the injection of both impactor- and target-derived sulfur dioxide into the stratosphere, and emphasized that 10,000-year events may directly deposit enough sulfur dioxide to surpass the normal stratospheric sulfate burden of about 10^{12} g by a factor of 5 or more.

The third important aspect of impactor composition is the total content of alkali metals, alkaline earths, and transition metals, which strongly enhance the radiant efficiency of the vapor produced during ablation. Radiant efficiencies for impacts have been estimated to be as low as 0.01% and as high as about 30% (Svetsov, 1996). However, the latter estimate considers only shocked-gas opacity due to the chemistry of air, in which, at high temperatures, moderate opacity can be contributed by very minor species such as the CN radical. One might expect that additional sources of opacity from rock vapors such as those listed previously would cause unity optical depth in the plasma sheath to occur farther from the surface of the ablating body, where the temperature is lower than in Svetsov's treatment, but the effective radiating area is larger. Because the luminosity of the dense, shocked gas varies roughly with the temperature of the radiating "photosphere" raised to the fourth power, it would not be surprising to find radiant efficiencies rather lower than 0.3. The detailed model calculations of Nemtchinov *et al.* (1997) described in Chapter 2 suggest that more typical luminous efficiencies would be about 0.15 for high-speed entering rocky and metallic objects, but that the luminous efficiency may in some cases actually approach 0.30.

Finally, the water content of the impactor is important in some settings, such as modeling of impacts into the atmosphere of Venus, but is of little significance for the present application of the impact simulation program. Water content enters into terrestrial impact phenomena through the catalytic effect of water photolysis products (OH and OOH) in destroying atmospheric ozone, and through providing an oxidizing and hydrating agent for the production of sulfate areosols.

For the purposes of this simulation, no compositional distinction is made between long-period and short-period comets. Both are taken to be 60% water ice, 10% other HCNO volatiles, and 30% carbonaceous chondrite (CI) residue.

Physical Properties of Impactors

Several of the physical properties of entering bodies are of great importance. The principal properties of interest are itemized next.

DENSITY

The bulk density is well established for a wide variety of meteorite types. The main uncertainty lies in the way in which porosity is considered in poorly consolidated solids. The densities of comet nuclei are only very poorly known, probably lying in the range of 0.2–1.0 g cm^{-3} in most cases. Even porosity measurements on low-density meteorites that are available in the laboratory are virtually nonexistent because of fears of contamination of these rare, valuable specimens by the fluid used to measure their pore volume. For macroscopic asteroidal bodies, densities are taken to be the same as in laboratory samples of meteorites of the same compositional class. The densities of the most nickel-rich iron meteorites range from 7.5 up to nearly 8 g cm^{-3}, and those of the most friable carbonaceous meteorites are close to 2.4 g cm^{-3}. Typical stony chondritic meteorites have densities of about 3.7 g cm^{-3}.

STRENGTH

The crushing strength is of fundamental importance in modeling the aerodynamic breakup of high-speed bodies. The point of departure for determining crushing strengths is laboratory measurements on centimeter-sized samples of meteorite materials (Table 3.5). We do not, however, assume that these laboratory-scale strengths apply directly to kilometer-sized bodies. In fact, large solid bodies contain cracks on a wide variety of scales. The larger the body, the wider the range of size scales it contains, and the more cracks it contains per unit volume. Thus the crushing strength must decline steadily with increasing size for compositionally and structurally uniform material. This behavior has long been recognized by materials scientists. The simplest semiempirical treatment of this phenomenon, the Weibull strength law, is in the form of a power law:

$$S = S_o (m/m_o)^n, \tag{3.2}$$

as illustrated in Fig. 3.1. Here the subscript o denotes laboratory-

TABLE 3.5

Measured Strengths of Impactors

Class	Strength (dyn cm^{-2})
Lunar regolith	0
Lunar basalts	2.5×10^9
Irons	
Room temperature (ductile)	3.6×10^9
Low temperature (brittle)	?
Stones (L chondrites)	0.06×10^9 to 2.6×10^9
Fireballs that drop stones (Lost City, Innisfree)	$>0.2 \times 10^9$
Tunguska projectile	0.2×10^9
Fireballs	
PN40503	0.07×10^9
EN160166	0.05×10^9
Cometary fireballs	10^5–10^6

scale sample masses (1 g) and strengths. Note that the maximum strengths of meteorite materials are about 3.5×10^9 dyn cm^{-3} (3.5 kbar), appropriate for both iron meteorites and pallasitic stony irons, which have a continuous metallic matrix and disintegrate at the failure point of metal. The weakest meteorites, the Ivuna-type carbonaceous chondrites, have laboratory crushing strengths of only 1–10 bars. The somewhat less volatile-rich CM (Murchison-type) carbonaceous chondrites are roughly 10 times stronger. For stones, the exponent n is -12.0. The variable m_o is the mass of the sample on which the strength is measured in the laboratory, typically on the order of 1 g.

The strengths of iron bodies are complicated by the existence of a brittle–ductile transition that occurs at temperatures

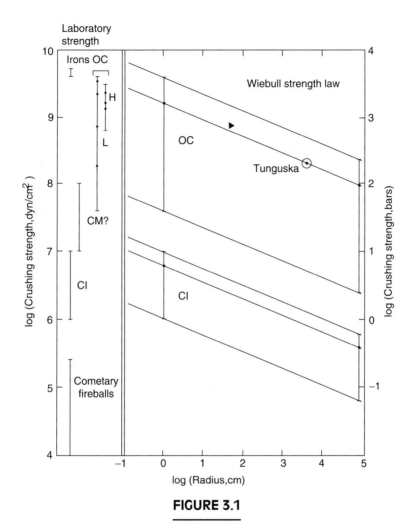

FIGURE 3.1

Laboratory and dynamic crushing-strength data. The centimeter-scale properties are extrapolated to larger sizes using a Weibull strength law. The Tunguska object appears to be a typical ordinary chondritic stone, as first argued by Sekanina (1983).

reasonable for asteroidal solids. The transition temperature T_{tr} for Ni-rich metal is quite low, about 125 K for metal containing 20% Ni and 1.5% Co, and about 450 K for metal with only 6% Ni (Remo, 1994). The data show that $1/T_{tr}$ varies linearly with the percentage Ni content f_{Ni}:

$$T_{tr} = 1000/(0.444 f_{Ni} - 0.405). \qquad (3.3)$$

For typical mean temperatures in the inner belt, about 150 K, most irons (all those having less than 16% Ni) would be brittle. For small iron bodies in orbit near Earth, however, with mean temperatures of about 250 K, only those irons containing less than 10% Ni would be brittle. A histogram of Ni contents of irons and stony-irons, drawn largely from the tabular compilation by Wasson (1974), is given in Fig. 3.2. Note that the hexahedrite irons (members of Wasson's IIA chemical group), with 5.25–5.90 wt % Ni, dominated by the low-Ni alloy kamacite, would clearly be brittle. Likewise, most octahedrites would be brittle. The IIB irons (5.5–6.5% Ni, coarsest octahedrites), the IA, IIE, IIIE, and IIIF irons (6–9.7% Ni; coarse octahedrites), the IIIA irons (7–9.3% Ni; medium octahedrites), and the IVA irons (7.3–10% Ni; fine octahedrites), which together account for the large majority of all irons, should all be brittle upon encounter with Earth. Several groups are transitional, including the IIC (9–11.5% Ni), IID (9.6–11.3% Ni), and IIIB (8–10.5% Ni) irons, which are mostly medium octahedrites. Predominantly ductile classes include IB (9–25% Ni) and IIIC (10.5–13% Ni) irons. The IIID (16.6–23.6% Ni) and IVB (15.8–18% Ni) irons, which are dominantly nickel-rich ataxites, are assuredly ductile. Among meter-sized and smaller irons, which rapidly reach thermal steady state with the solar radiation field as they traverse their orbits, these considerations favor the preferential survival

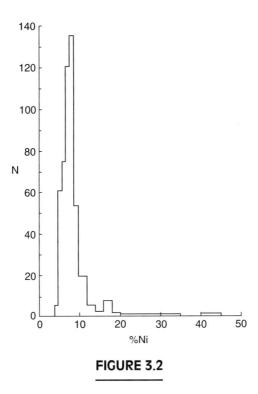

FIGURE 3.2

Histogram of nickel contents of iron meteorites. (Data from Wasson, 1974.)

of the individuals irons with the highest nickel contents, a phenomenon already noted in the meteorite literature. For 100-m or larger iron bodies, whose deep internal temperatures respond only slightly to changing insolation, the characteristic temperatures in their interiors reflect the time-averaged insolation they receive as they orbit the Sun, which lends extra weight to the low temperatures experienced while the asteroid is moving slowly through aphelion. Such metallic asteroids would therefore be brittle almost irrespective of their composition. However, such bodies are so dense and massive that they can penetrate deep

TABLE 3.6
Physico-chemical Properties of Impactors

Class	Density (g cm^{-3})	Strength[a] (dyn cm^{-2})	W (H$_2$O)	τ[b]	ΔH (erg g^{-1})
Irons	7.90	3.7×10^9	0	0.3	8.08×10^{10}
Stony-Irons					
Mesosiderites	5.00	3.0×10^9	0	0.3	8.08×10^{10}
Pallasites	3.3–5.3	3.0×10^9	0	0.3	8.08×10^{10}
Achondrites	3.20	1.7×10^9	0	0.02	8.08×10^{10}
Chondrites					
Ordinary H, L, LL	3.6	0.7×10^9	0	0.3	8.08×10^{10}
Carbonaceous CM	2.75	0.05×10^9	0.10	0.05	8.08×10^{10}
Carbonaceous CI	2.25	0.01×10^9	0.20	0.05	8.08×10^{10}
Comets					
Short period	1.40	0.1×10^9	0.50	0.02	3.2×10^{10}
Long period	1.10	0.1×10^9	0.75	0.02	3.2×10^{10}

[a] Crushing strength on 1-cm scale.
[b] Luminous efficiency.

into the atmosphere before disruption. The internal thermal state of metallic impactors is probably irrelevant at sizes larger than about 100 m.

Mesosiderites, with 6–10% Ni, and pallasites, with 8–11% Ni, are marginal, but most are probably brittle. Note that the crushing strength of pallasites is almost certainly dominated by the continuous metal matrix, whereas that of mesosiderites should be dominated by brittle silicates, and hence relatively insensitive to the state of their included metal grains.

ABLATION RATE

The rate of removal of surface material by ablation is controlled by a complex combination of factors, including heat capacity, thermal conductivity, melting point, magma viscosity, and so on. The net effect of all these influences has, since the time of Öpik, been modeled as a single number for each material, the ablation coefficient.

LUMINOUS EFFICIENCY

The visible luminosity of a fireball represents only a small percentage of the total rate of energy dissipation by the body. The ratio of luminous energy emission to total energy dissipation rate, the luminous efficiency, depends on the temperature of the wake gas (and hence on the entry velocity), as well as the abundance of vapor components that have rich emission spectra at visible and ultraviolet wavelengths. The latter species are, as mentioned earlier, mostly transition metals and alkali metals.

Table 3.6 summarizes the compositional and physical properties of the different classes of bodies recognized in the simulation. These figures are strongly influenced by the work of Nemtchinov *et al.* (1997).

4

The Impact Process

For each impact event, the mass and physicochemical properties of the body are specified. A heliocentric orbit is selected from an envelope of orbital parameters appropriate to bodies of this composition class and mass, resulting in an encounter velocity with Earth. For our purposes, bodies large enough to represent a hazard to Earth experience no significant atmospheric interactions at altitudes above about 100 km. We treat the orbits of incoming bodies by first calculating their velocity at an altitude of 140 km. This is done simply by adding the gravitational potential energy difference between 140 km and infinity to the free-space encounter velocity appropriate to the body's orbit about the Sun.

By conservation of energy, the encounter velocity is thus corrected to an entry velocity at 140 km altitude. The initial entry angle H_0 (in degrees) is generated randomly in accordance with a long-established and simple theory (see, for example, Shoemaker, 1962):

$$Q = 2 \text{ Rnd} - 1, \tag{4.1}$$
$$\Theta_0 = 28.6479 \, (1.5708 - \sin^{-1} Q). \tag{4.2}$$

Such an entry-angle model features near-zero probabilities of entry close to 90° (vertical incidence) and 0° (grazing incidence), with the most probable entry angle being 45°. The entry trajectory is followed from an altitude of 140 km (for Earth) until one of the following fates has been realized: burnup, catastrophic breakup, impact with the surface, or skipout through 140 km altitude. Below 140 km the acceleration of the body by Earth's gravity is explicitly taken into account.

Atmospheric skipout with capture by aerodynamic drag was first discussed by Baker (1958). Bodies on skipout trajectories have been reported by Jacchia (1974) and by Borovicka and Ceplecha (1992). Roughly 1% of the impacting bodies, generally those with entry angles below 9°, skip out successfully from perigees below 140 km.

The model atmosphere used to calculate the behavior of entering bodies is a smoothed exponential decrease of density ρ above a spherical surface:

$$\rho_{atm} = \rho_0 \exp(-z/H), \tag{4.3}$$

where H is the mean scale height $RT/\mu g$, T is the mean temperature of the atmosphere below 140 km, and μ is the mean molecular weight.

We do not for any purpose assume a plane-parallel atmospheric model, nor do we use only one or two fixed entry angles. The atmospheric model is therefore appropriate for calculating the fate of bodies with all possible entry geometries, including low incidence angles. The geometry of the model permits accurate modeling of both atmospheric skipout and aerobraking capture into temporary low-perigee Earth orbits.

The aerodynamic deceleration of the body is calculated from its size, velocity, air density, and drag coefficient:

$$dV/dt = -0.5 C_D \rho V^2 A/m. \qquad (4.4)$$

The acceleration produced by change in altitude (gravitational potential energy) is

$$dV/dt = (g \sin \Theta)/m. \qquad (4.5)$$

[Note that Eq. 1 of Chyba *et al.* (1993) contains an error in that it equates the gravitational force $m\, dV/dt$ rather than the acceleration dV/dt to $(g \sin \Theta)/m$. The calculations he reports have been carried out with the correct equation.] The trajectory is numerically integrated to take into account both aerodynamic deceleration and gravitational acceleration, resulting in trajectory bending. In general, of course, the trajectory is always concave downward in spherical coordinates, but the rate of change of altitude may reverse sign for bodies with very shallow entry angles. Such bodies almost always successfully skip out of the atmosphere. Although the model permits the much rarer skip-glide entry trajectory as well, it has never been observed in actual runs, in large part because tracking of the object is terminated once it rises through the 140-km altitude level.

Aerodynamic lift forces are assumed to be both small and random in direction, since only the smallest entering bodies (those too small to be important hazards) can achieve stable orientation during the brief time of their interaction with the atmosphere. Fragmenting objects, which rapidly change their geometry, should satisfy the condition that their mean lift coefficient be close to zero. Chyba *et al.* (1993) have presented the results of some simulation calculations that bear out the negligible effect of aerodynamic lift.

Interaction with the atmosphere has a number of other important effects besides deceleration. These include luminous emission of energy by the fireball as it passes through the atmosphere, ablation, fragmentation and dispersal of solids, terminal airburst detonation, radiant heating by the terminal flare, blast wave generation and propagation, ignition and suppression of surface fires by the two preceding effects, nitrogen oxide generation in the shocked gas, and injection of meteoritic materials such as dust and sulfur into the atmosphere.

The total rate of energy dissipation by the entering body is given by

$$dE = 0.5v^2 dm + mvdv. \qquad (4.6)$$

The ablation rate of the projectile is related to the entry velocity by

$$dm = (0.413 A \Gamma \rho_{atm} v^3 / H_{vap})\, dt, \qquad (4.7)$$

where Γ is the ablation coefficient, A is the geometrical cross-section area, and H_{vap} is the molar heat of vaporization of the projectile (Baldwin and Sheaffer, 1971; ReVelle, 1979).

The luminous flux from the fireball on Earth's surface directly below a fireball of altitude z is given by

$$F = -(\tau V^2/8\pi z^2)\, dm/dt, \qquad (4.8)$$

where τ is the luminous efficiency (Baldwin and Sheaffer, 1971; Saidov and Simek, 1989). Fireball observations both in North America and Europe provide a rich body of data on the luminous behavior of entering bodies (see, for example, McCrosky et al., 1971; Halliday et al., 1989). Very weak massive bodies are so luminous that they can ignite fires over areas of hundreds to tens of thousands of square kilometers (Nemtchinov and Svestov, 1991).

Fragmentation is modeled as occurring when the stagnation-point pressure exceeds the crushing strength of the projectile at its largest (intact) scale size. The fragments, having smaller scale sizes, are slightly stronger than the parent projectile. The larger fragments are crushed slightly lower in the atmosphere than the original projectile, breaking into yet smaller and slightly stronger fragments. The normal course of fragmentation then is for the incoming body to undergo a rapid series of breakups, culminating with fragments that either crush into unrecoverably fine powders, vaporize completely (Svetsov, 1996), or decelerate sufficiently so that they reach the ground at low speeds (below the speed of sound in their solid material) as intact, macroscopic pieces, which we call meteorites.

The acoustic signature of large entry events has been noted countless times in the literature. The sonic boom laid down on the ground by large entering bodies has even been used to calculate the trajectory of the entering body. One such event, a bolide with tens of kilotons of total energy, was tracked using data from seismic sensors deployed in the Hiroshima area (Nagasawa and Miura, 1987).

The shock wave generated by passage through the atmosphere and by the terminal explosion heats a volume of atmosphere to temperatures so high that nitrogen in the atmosphere is partially oxidized to NO via the reaction

$$N_2 + O_2 \rightarrow 2NO. \qquad (4.9)$$

The equilibrium NO concentration in the shocked gas is a function of both the total pressure and the peak temperature reached behind the shock (Park, 1978; Lewis et al., 1982; Prinn and Fegley, 1987). If cooling of the post-shock gas is rapid (i.e., if the scale size of the shocked gas region is of order 1 km or less, comparable to that in military nuclear explosions), then the high equilibrium NO abundances calculated for the shocked gas are also closely valid for the rapidly cooled, quenched gas as well. But for very large explosions (an atmospheric scale height or larger) the cooling of the post-shock gas is slow, allowing for kinetic reequilibration of the cooling gas, and thus reducing the NO concentration substantially (Zahnle, 1990). Zahnle's treatment suggests that the total number of moles of NO produced by an impact with energy Y megatons should be approximated by

$$\log(N_{NO}) = 8.5 + 0.8 \log Y. \qquad (4.10)$$

For the purposes of the present model, we will be dealing with 100- to 1000-year impact timescales, for which the magnitude of the NO production is small enough to render it of purely local significance. In that size range he finds that a production rate of about 2×10^8 moles per megaton is appropriate. The size of 100-year (and even 1000-year) events is within the range of experience with large nuclear explosions on Earth, permitting us to ignore the effects of reequilibration during slow cooling: We adopt a production efficiency of 1.5×10^{-13} NO per erg of dissipated energy. Nitrogen oxides can catalyze the destruction of ozone. Turco *et al.* (1981) have presented evidence for ozone depletion associated with the Tunguska event in 1908.

Further oxidation of the NO produced by blast waves makes NO_2 and its dimer N_2O_4, which, on hydration

$$N_2O_4 + H_2O \rightarrow HNO_2 + HNO_3, \qquad (4.11)$$

produces both nitrous and nitric acids, which are highly water soluble and rain out as acid rain. Evidence for acid-enhanced continental erosion immediately above the *K-T* boundary has been presented in MacDougall's (1988) study of abrupt changes in the strontium isotopic composition in post-Chicxulub marine sediments. These effects are, however, dominated by the fortuitous targeting of a calcite- and anhydrite-rich sedimentary deposit, which generates an acid burden that owes far more to sulfur oxides from anhydrite destruction than to nitrogen oxides from the atmospheric blast wave. Crutzen (1987) also points out that the enormous soot content of the *K-T* boundary layer implies wildfires on such a scale that vast amounts of nitrogen oxides would be produced along with the pyrotoxins and carbon oxides that are always associated with natural biomass burning.

Nitrous acid, a major product of nitrogen tetroxide hydration, is of special interest because it is a carcinogen, teratogen, and mutagen.

Finally, atmospheric explosions of fireballs inject the component materials of the projectile into the atmosphere, often at altitudes above 20 km. The principal materials injected into the stratosphere are dust and sulfur compounds. Meteoric sulfur injection on a small scale has been reported by Rietmeijer (1989). Total vaporization of most common classes of meteoritic material in air results in the generation of large amounts of SO_2. A 3×10^{12} g carbonaceous impactor (a 1000-year event; roughly 100 Mt TNT) contains about 2×10^{11} g sulfur, which is oxidized in the cooling explosion fireball to 4×10^{11} g SO_2, and on further oxidation and hydration, to 6×10^{11} g of H_2SO_4. The impact injection of other volatiles, while sometimes important (especially on Venus), is of lesser importance on Earth. Such injections, however, including the deposition profiles of meteoritic halogens in the stratosphere, have not been well studied. Kring *et al.* (1996) show, however, that high-altitude airbursts (at 30 km or higher) are not required for massive sulfur injection; much larger explosions at lower altitudes generate hot plumes that lift the shocked gases and projectile vapor to high altitudes.

Weak and fast-moving bodies experience aerodynamic forces that greatly exceed the crushing strength of their materials while still well above the ground. Such bodies deposit a large fraction (50–99%) of their total energy content over path lengths much smaller than an atmospheric scale height, effectively as a point-source energetic explosion. Such yields are normally reported in units of megatons (Mt) of TNT. Experience with atmospheric testing of nuclear weapons permits calculation of the blast overpressures experienced on the ground as a function of impactor explosive yield, burst height, and slant range from the explosion (Glasstone and Dolan, 1977). Several regimes of behavior are known. First, if the blast is high enough, overpressures in excess of about 2.5×10^5 dyn cm^{-2} (0.25 atm) will not be experienced on the ground. Although damage and injuries may result, such blasts are incapable of causing structural failure of normal-strength construction. (Failure of very low quality structures usually involves so little mass that building collapse is not likely to have lethal consequences). The critical blast altitude at which the

0.25-atm overpressure surface just reaches the ground is taken to be the maximum blast height for lethality. Explosions very close to or on the surface shock a small region extremely intensely, but actually devastate a smaller area (i.e., subject it to 0.25-atm overpressures) than an airburst with the same yield. There is, in fact, an optimum altitude for an airburst of given yield, at which it subjects the greatest possible area to building-leveling blast effects (Fig. 4.1).

Bodies that have a high enough strength and low enough velocity will often survive passage through the lower atmosphere and impact the surface at nearly the same speed with which they entered the atmosphere. Bodies that begin to disrupt only within 30 to 100 body diameters of the ground strike the surface of Earth as compact debris clouds that have virtually the same effect as unfragmented projectiles. Also, sufficiently massive impactors, almost irrespective of their strength, will successfully penetrate the atmosphere and strike the surface at very high speeds. On a 1000-year timescale, the bodies that reach the surface intact (or nearly so) are mostly rare bodies of exceptional physical strength, especially irons and pallasitic stony-irons. Mesosiderites and achondrites, and some of the most extensively recrystallized chondrites, are several times weaker than the strongest irons, but can on occasion enter with both low speed and shallow entry angles, allowing some of them to approach the surface closely. On the 1000-year timescale, carbonaceous and cometary impactors almost never succeed in penetrating deeply enough into the atmosphere to impact the surface (Chyba, 1993).

The first parameter of interest for a body that penetrates into the lowermost 1 km or so of the atmosphere is its total energy content. The second factor that governs solid-surface impact behavior is the effective density of the impactor. For those

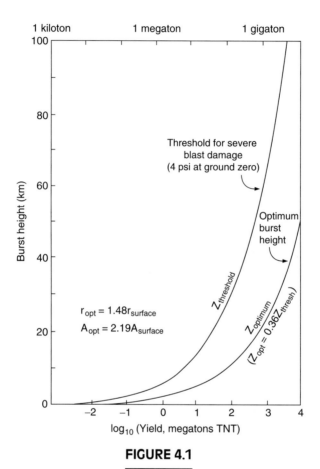

FIGURE 4.1

Regimes of surface effects versus burst height and explosive yield. The curves are the maximum altitude for destruction of normal buildings on the surface (the 4-psi overpressure surface) and the optimum burst height, defined as that height at which the surface area subjected to dynamic pressures greater than 4 psi (0.25×10^6 dyn cm^{-2}) is maximized. The contours for 1-psi overpressure, sufficient to shatter 50% of all plate-glass windows and accelerate the fragments to speeds of more than 100 m s^{-1}, lie somewhat higher.

that reach the surface of Earth intact, that is simply equal to the impactor density. But for debris clouds, the effective density is significantly reduced by entrained shock-compressed air.

When the stagnation-point dynamic pressure exceeds the (size-dependent) crushing strength of the impactor, the body breaks up into a swarm of fragments The dispersal velocity of this swarm is given by

$$V_{disp} = 1.9 V(\rho_{atm}/\rho), \qquad (4.12)$$

where ρ_{atm} is the ambient atmospheric density and ρ is the density of the projectile. The radius of the swarm increases at this rate, causing the effective frontal area to increase more rapidly than t^2. Once fragmentation begins, the effective crushing strength of the fragments is increased in accordance with the Weibull law. The result is a series of disruptions occurring in close sequence, made more complex by the greatly enhanced deceleration and ablation of the entering body. Some bodies reach the ground as compact, hypersonic debris swarms. Others, especially exceptionally strong or slow bodies and those too large to disperse significantly before reaching the ground, will impact with speeds of many kilometers per second, excavating transient craters. The serial fragmentation and impact of the Campo del Cielo iron is discussed by Cassidy *et al.* (1965). The formation of multiple craters was explored by Passey and Melosh (1980), and the Pampas crater chain was investigated by Schultz and Beatty (1992) and by Schultz and Lianza (1992).

Cratering behavior is modeled as in Melosh (1989). The formation of a transient crater, release of dust and vapor, high-speed ejection of debris, collapse of the transient crater, and central peak formation are all extensively discussed in the literature, but of slight relevance to the present study.

Dust raised by large surface impacts can have devastating climatological consequences. The mass of dust lifted to stratospheric elevations by surface impacts is modeled according to the approach of Toon *et al.* (1995). We here include impact vapor and melt along with dust, with an area loading of globally distributed fine material given by

$$s\,(\text{g cm}^{-2}) = 10^{-8} \log Y\,(\text{Mt}), \quad (4.13)$$

where Y is the explosive yield in megatons of TNT. For large impacts, the climatological effects of stratospheric dust injection may be very severe (Turco *et al.* 1983). However, on the timescales of interest in the present study, events are generally too small to elevate significant masses of dust. All impactors, however, deposit a large fraction of their own total mass in the form of atmospheric dust. The chemical signature of this extraterrestrial material, as noted earlier, is usually detectable as a dramatically higher content of platinum-group metals, most notably iridium. Again, the iridium signatures of events of interest to us on the century or millennium timescale are usually only local.

The seismic effects of impacts are often mentioned, but the efficiency of seismic wave generation is often assumed to be very high. Actually, coupling between a typical impact and the surface is inefficient, and earthquakes with moment magnitudes up to about 9 are surely more common than impact-driven earthquakes of similar magnitude. Kisslinger (1992) has argued that the largest plausible earthquake is near moment magnitude 9.0. Figure 4.1 compares the observed distribution of earthquake magnitudes to the total energy content of impactors (i.e., assuming perfect coupling of impactor kinetic energy to crustal displacement). It appears that impactor yields well beyond 10^5 Mt would be required to surpass the magnitude of the largest possible endogenic earthquake. Such events have mean occurrence intervals of several hundred thousand years (impactor diameters of 1 to 2 km). On million-year or longer timescales, the largest seismic events that can occur on Earth are due to impacts.

Energetic (gigaton and larger) impacts into deep water (kilometers deep) excavate a transient crater that may, in the most extreme cases, reach and erode the abyssal ocean floor. The conical spray sheet can easily exit the atmosphere, and superheated water vapor from the fireball can splash over intercontinental distances, devastating the ozone layer. Collapse of the transient crater in a convergent wavefront causes formation of a towering pillar of water comparable in height to the atmosphere, which then in turn collapses and rebounds, generating a tsunami wave train (Streliz, 1979; Croft, 1982; McKinnon, 1982; Melosh, 1982; Ahrens and O'Keefe, 1983; Sonett et al., 1991; Hills et al., 1994). These considerations, combined with certain semiempirical rules presented by Glasstone and Dolan (1977) based on experience with tidal wave generation by underwater nuclear explosions, can be summarized by

$$h\,(\mathrm{m}) = 135\,Y^{0.54}/r, \qquad (4.14)$$

where h is the peak wave height above mean sea level, Y is the explosion yield in megatons TNT, and r is the distance from ground zero in kilometers. The propagation speed of the wave is

$$V = (gd)^{1/2}, \qquad (4.15)$$

where d is the depth of the ocean. This is essentially identical to the treatment of impact-induced tsunamis by Hills and Goda (1993). It tacitly assumes that the entire tsunami occurs in deep water; that is, the depth of the water is several times the depth of the transient impact cavity.

Toon *et al.* (1995) have improved the physical plausibility of this tsunami model by using the laboratory data of Gault and Sonett (1982) to define the crossover between shallow-water and deep-water behavior. Essentially, in the shallow-water regime the distance to which a wave can propagate with a given height (for example, 10 m) is proportional to the fourth root of the explosive yield, whereas the distance it can reach with that same height in deep water is roughly proportional to the square root of the explosive yield (the 0.54 power, according to Eq. 4.14). The Hills and Goda treatment therefore greatly overestimates the effect of those impacts that are large enough, and excavate deeply enough, so that the shallow-water approximation ought to be used. Their approximation begins to diverge strikingly at a yield of several gigatons, and becomes ever less reliable toward higher yields. Although only one event in the present simulation (number 3005-354) lies outside the range of applicability of the deep-water approximation, care must be exercised when comparing calculations reported for longer time spans. For example, teraton (1 Tt = 10^6 Mt) impacts into deep water, events with mean recurrence intervals of a million years or more, can generate megatsunamis, with open-ocean heights on the order of 25 m (at a range of 5000 km) and runup heights of about 750 m (see Eq. 4.14). The model of Toon *et al.* (1995) implies an open-ocean wave height of about 10 m at a range of only 3000 km.

Toon *et al.* (1995) define the boundary between deep-water and shallow-water models as occurring at the lesser of 0.45 $Y^{0.25}$ and 0.3d. The former is a scale set by the radius of the transient impact cavity, and the latter is the depth scale of the ocean (d is the water depth in the target region). For example, for 4- and 6-km depths, all explosions with yields less than 50 and 250 Mt will be in the deep-water regime, and all larger explosions should be treated as shallow-water impacts. Their suggested formula for the area affected by tsunamis higher than h km in the deep-water regime is $A(h) = 0.29Y/h^2$ (which applies to yields lower than $0.2d^4$) and $A(h) = 0.13d^2Y0.5/h^2$ (which applies to yields higher than $0.2d^4$). This approach predicts more severe tsunamis for smaller impacts, but less severe tsunamis for the largest ocean impactors, in the gigaton range and higher.

One such enormous tsunami event has been documented. Gersonde *et al.* (1997) have reported on their studies of an asteroid impact feature on the floor of the Bellingshausen Sea in the Southern Ocean, in 5000 m of water, with an age of 2.15 Ma and an explosive yield of roughly 1 Tt. The impacting body, a mesosiderite, has been identified from large recovered fragments found in giant-piston core samples of the ocean floor. The impacting body was roughly 2 km in diameter, although that figure remains uncertain by about a factor of 2.

Very large impacts, well beyond the range encountered in this simulation, are capable of eroding atmosphere from the target planet. This effect was proposed as an important agent in the evolution of the atmosphere of Mars by Watkins and Lewis (1986) and modeled in detail by Melosh and Vickery (1989). The general theory for explosive blowoff from the terrestrial planets is treated by Vickery and Melosh (1990).

5

The Target

Population Density and Distribution

The method of assessment of the effects of impact bombardment on human populations is drawn heavily from the nuclear weapons literature. All the principal effects of nuclear weapons except those directly linked with radioactivity are germane.

Because this simulation is concerned with both the mean fatality rate expected from impacts and the intrinsic statistical variability of this fatality rate, it is inappropriate to use a population model that assumes a uniform distribution of people over Earth. Instead, the surface of Earth is divided into two general zones, solid surface and water. For solid land, population data compiled from a wide variety of encyclopedia, map, and almanac sources were used to generate a curve for the population density distribution of Earth's land areas. The population density on land was generated by:

$$\log \Sigma_{land} = 10^{0.1737 \ln(\ln(Rnd))} \qquad (5.1)$$

where Σ is the population density (km^{-2}).

For the oceans, data on the population density distribution are not directly available because it is compounded of factors that include the number, location, and occupancy of military, freight, and passenger ships, commercial and military airline traffic, and fishing vessels. An estimate of the number of people on and above the oceans was generated. This distribution, like that of population density on land, displays a high level of clustering because of the fact that most ocean and aviation traffic is confined to a finite number of densely populated vehicles plying a set of air traffic corridors, sea lanes, and fishing grounds that leave most of the oceans empty of human presence. In the absence of an enormous body of detailed data, the population density distribution over the ocean was assumed to be given by:

$$\log \Sigma_{ocean} = 8 \, Rnd^4 - 9. \qquad (5.2)$$

The growth of global population has been accompanied by a disproportionate increase in the population density on seacoasts. Seacoast urbanization has been accompanied by the concentration of national capitals on the sea or close to sea level. This global trend is especially noticeable in West Africa. Thus tsunamis strike target areas that have at least as high a population density as the rest of Earth's land area.

Hazards to Populated Areas

The hazards to which the population are exposed include blast-wave acceleration of window glass, radiant ignition of fires by the fireball and terminal explosion of the impactor (radiative fluences in excess of 10^9 erg cm^{-2}), structural failure of buildings due to high blast overpressures ($P_{dyn} > 4$ psi $= 2.5 \times 10^5$ dyn cm^{-2}), eardrum rupture ($P_{dyn} > 20$ psi $= 13 \times 10^5$ dyn cm^{-2}), and infliction of fatal injuries by being tumbled by the blast wave ($P_{dyn} > 4$ psi $= 2.5 \times 10^5$ dyn cm^{-2}).

When radiative ignition of fires occurs in an urban area, conditions conducive to ignition and propagation of firestorms are likely. Thresholds for ignition of flammable materials, including window drapes and shades, upholstery, clothing, trees and bushes are taken from Glasstone and Dolan (1977).

Hazards in and on the Ocean

Large impacts in the ocean can generate tsunami waves that have destructive power out of proportion to their total energy. Conceptually, airbursts disperse their energy over the growing surface of a sphere, so that, in a gross sense, the radiative and dynamic effects of the blast drop off with the square of the distance from the explosion site. (The more complex but less common case of explosions whose scale sizes or altitudes are comparable to or greater than the atmospheric density scale height is fully considered in the model.) An ocean surface blast, however, excavates a large transient crater in the ocean, which then relaxes, implodes to generate a towering pillar of water at its center, relaxes again, and pulsates in this manner for several cycles. These pulsations generate a train of waves that travel radially outward from the explosion site, as discussed in Chapter 4. Typical deep-water wave velocities, given by Eq. 4.15, are about 900 km hr^{-1}. The profile of the wave features a wavelength of several kilometers (increasing with the scale size of the transient crater) and modest heights, often of order 1 m even for waves that later do considerable damage. The energy carried by the waves is spread about the periphery of a circle, and therefore decreases in intensity with $1/r$ instead of the $1/r^2$ characteristic of explosions in the atmosphere. Even making allowance for phase disturbances in the wave caused by depth variations in the ocean, refraction and interference about islands in island chains, and energy loss by waves breaking on islands and atolls, the intensity of tsunami waves drops off only a little more rapidly than $1/r$. At very large distances, convergence of the wavefront toward the antipodal point can halt and even reverse this rate of decay of wave energy.

As the tsunami wave train encounters a continental margin, retardation of the leading (shallow-water) part of the wave causes a dramatic shortening of the wave and growth in its height. Such waves typically run up in height by a factor of 30, although the runup factor is so sensitive to local ocean-floor topography and coastline shape that a range of runup factors from about 10 to 100 is possible locally. The penetration of tsunami waves into low-lying land areas is treated as in Hills *et al.* (1994).

As a rule, bodies less than about 100 m in diameter usually break up in the atmosphere and are therefore ineffectual at generating tsunami waves. The few bodies of this size that are strong enough to penetrate intact to the ocean surface are usually irons, stony-irons, or hard-rock achondrites. These unusually strong impactors make up roughly 5% of the bombarding flux. Above the 100-m size range, the frequency of ocean impacts drops off more slowly with increasing size than the incident flux does. This happens because progressively higher proportions of the incident flux succeed in penetrating the atmosphere and impacting the ocean surface. At a size of about 2 km, almost all incident bodies survive to impact. No rule of thumb is used to estimate the percentage of surface impactors in a given size range; every individual entry event is numerically integrated to determine the fate of the projectile.

Property Damage and Destruction

The expected economic losses from impact events are discussed by Canavan *et al.* (1994) in the context of a statistically averaged impact flux model. Their differential loss curves for impacts consist of four straight-line segments on a log (loss) vs. log (mass) plot, with abrupt discontinuities of a factor of 20–50 between adjacent segments. An impact model that takes into account the full statistical variability of initial orbital parameters, entry conditions, impactor physical properties, and impact site, such as the present Monte Carlo model, does not exhibit such discontinuous behavior.

A more direct way to estimate losses is to realize that the geographical distribution of humans and of human assets are very similar; so much so that the ratio of fatalities to economic loss is arguably nearly a constant. Our approach, therefore, is to use human casualties as a proxy for economic loss. Note that this approach does not mean assigning a cash value to human life; rather, it is an estimate of the average cash value of property and goods destroyed per human death. Conversion of impact deaths to expected economic loss can be achieved to adequate accuracy by multiplying the number of deaths by a constant. A comparison of Canavan's cost analysis with the lethality predictions of Morrison *et al.* (1994) or with the present work suggests a conversion factor of about $100,000 per person. This factor in turn suggests that a mid-20th-century global population of 5×10^9 people would have a total property value of $\$5 \times 10^{14}$, which, at a gross global product level of about $\$2.5 \times 10^{13}$ per year, represents about 20 years of global product. This seems a reasonable estimate. Therefore an analysis that suggests an expected time-averaged fatality rate of 3000 people per year would likewise suggest an expected mean direct economic (property) loss of about $300 million per year, in accord with Canavan's estimate. A global insurance policy against impacts with premiums of $100 million a year would be an excellent bargain. Over the long term, an insurance company offering such coverage would have to charge premiums of at least $400 million per year to remain in business. Thus any combination of prevention and remediation costs (discover, tracking, characterization, interception, and diversion) costing $100 million would pay back that investment several times over. We shall return to this issue in Chapter 8. It should be amply evident, however, that a true global catastrophe that prevents agriculture for a year or otherwise destroys the infrastructure of civilization (including all insurance companies) is not an acceptable outcome no matter how much insurance we may carry.

Particularly Hazardous Targets

Four categories of targets that present special hazards seem worthy of mention. These are sensitive national security sites, sites with hazardous materials that might be broadcast by the impact, geologically sensitive sites such as volcanic and earthquake regions, and areas with essential roles in food production, storage, or distribution.

An impact that destroys a national capitol or command center may contribute disproportionately to chaos and hinder appropriate responses to the disaster. The "decapitation" of a highly centralized nation by loss of its government and command structure could seriously jeopardize its ability to regroup and render aid to outlying areas. In some cases, which seem ever less likely as communications and scientific understanding of impacts improve, the sudden unexpected loss of a militarily, politically, religiously, or culturally prominent site may trigger "knee-jerk" responses such as rioting or military action. For example, a major unexplained explosion in or near a military base or national capitol at the height of the Cold War could easily have triggered a nuclear exchange before calmer heads could prevail. Knowledge and good communications are the best antidotes to such behavior.

Impacts may in some cases cause release of hazardous materials into Earth's atmosphere. Sites with such hazardous materials would include nuclear power plants, nuclear weapons assembly or storage sites, munition plants and depots stocking chemical or biological warfare agents, chemical plants with large burdens of toxic materials, and refineries, oil fields, and petroleum tank farms. Sources of very large amounts of soot, or of chlorine and bromine, would be especially hazardous.

Triggering of earthquakes, volcanoes, or landslides requires events of such severity that the seismic responses become little more than ancillary details of the greater problem. Even if an impact of, say, moment magnitude 7.5 (a 100,000-year event) triggers an earthquake of magnitude 9.0, that earthquake is one that would have happened in any event as a natural consequence of the stress buildup that made it possible. The impact merely changes the timing of the earthquake. Indeed, by triggering it earlier, the impact actually diminishes the expected moment magnitude. Thus the devastation done by a triggered temblor would be no greater, and probably less, than that which would have otherwise occurred. The negative effects assignable uniquely to the impact, above and beyond the far greater damage done directly by the impactor explosion, would consist of the destruction and demoralization of disaster response personnel and equipment.

Impacts in sulfate-rich terrains were discussed in Chapter 3 as a worst-case extension of sulfur injection by meteoritic or cometary bodies.

Destruction of assets essential for food production, processing, storage, or distribution also presents a threat of starvation and social disorder. However, an international response to food emergencies is possible and effective for even a regional-scale disaster (a 10,000-year event). For the very largest (million-year or larger) impacts, disruption of food production is arguably the most devastating single consequence. However, on the timescale covered by this simulation, interference with food supplies is only a marginal threat.

In sum, it is good to be cognizant of these special hazards, but difficult to assess them quantitatively. No special allowance for them has been included in the simulation program. However, these considerations underline the principle that it is uncommon outliers of the impactor population that dominate the hazard statistics on 10,000-year and shorter timescales.

6

Method of Calculation

Program Structure

A guide to the details of operation of the program is provided in Appendix A. Here we treat only the general workings of the program and its subroutines.

The calculations reported herein were performed as a Monte Carlo simulation of the bombardment process. The time span under consideration, t, is divided into 100 or 200 steps of duration dt. In most of the results presented here, $t = 100$ or 200 years and $dt = 1$ year, except where some different time interval or duration is specified. The program first initializes all conditions (subroutine INITIALIZE) and sets up the headings of the table that will be produced as output (subroutine FORMPRINT). Only the largest (most energetic) single event in each time step is reported. The program begins by using the shape of the overall mass spectrum and a random number generator (subroutine MASS) to specify the mass of the largest body of the time step (usually 1 year). That body is next randomly assigned to either the asteroid or comet class, based on the statistical weights appropriate to a body of that mass, and then randomly categorized into a compositional and orbital family (subroutine KIND) based on the taxonomic statistics given in Chapter 3. Those bodies selected as asteroids can become irons, pallasites, mesosiderites, achondrites, ordinary chondrites, CM chondrites, or CI chondrites. Cometary bodies similarly become either short-period or long-period comets. The treatment of meter-sized cometary dustballs, while important on the daily timescale, is largely irrelevant with 1-year time steps, because such bodies are never the biggest event of the year, and of course never inflict injury or property damage.

Once the compositional and orbital family of the body is set, the encounter velocity is generated randomly from a velocity-frequency function that has the same envelope as observed for that class of body (subroutine ENERGY). The velocity distributions for the two groups of comets are well determined, as are those for stony meteorites. The carbonaceous meteorites are assigned to orbits that are, on average, more eccentric than those of the ordinary chondrites. The relatively uncommon igneous differentiated classes (irons, stony-irons, and achondrites), for which we do not have any good statistical sense of their range of orbital elements, are assigned to the same distribution as the ordinary chondrites. The velocity distribution determines the impactor kinetic energy density at entry. This subroutine also randomly generates an entry angle.

The mass and energy are then used to screen the incoming flux for objects capable of accelerating atmospheric gases to escape velocity (subroutine BLOWOFF). This feature was written for applications of the program to Mars, but is included in the Earth runs by default. Atmospheric erosion events on Earth require such high threshold energies that they are not encountered in the short runs reported here.

The atmospheric flight of the entering body is numerically integrated within a subroutine that also calculates the rate of energy loss, luminosity, ablation rate, deceleration, and fragmentation of the body (subroutine FRAGMENT). Outcomes discovered during numerical integration are flagged (burnup, deceleration to <3000 m s^{-1}, fragmentation, continental impact, ocean impact, skipout to escape, skipout to temporary orbit).

Those small bodies that decelerate to less than 3000 m s^{-1} are essentially classical meteorite falls and are not followed to impact. Experience with runs in which these meteorite falls are integrated down to the surface suggests that, although small meteorites may strike several people per century, such events cannot be responsible for a significant number of deaths. Those meteorites that strike the surface at a speed in excess of sound speed in soft rock (which is the significance of 3000 m s^{-1}) explode on impact and excavate a crater whose dimensions are calculated in subroutine CRATER. The formation of large impact craters, on the scale of Meteor Crater or an urban area, is sufficiently rare on human timescales that cratering phenomena are generally unimportant in simulations that cover timescales of a thousand years or less.

Hazards to life and property are treated separately (subroutine HAZARD). The program checks to see whether the planet is Earth; if not, it returns to the main program. The NO$_x$ mass produced by the fireball and terminal explosion is calculated and reported as a globally averaged concentration. Again, nitrogen oxide production in events on a 1000-year timescale is generally no more than a local inconvenience. A random number generator is used to assign the impactor to land or sea, and the population density distribution rule appropriate to that setting is then used with another random number generator to assign a population density to the target region. (Assignment to random logitudes and latitudes with the appropriate population density, and hence to real points on Earth's surface, can be done with an auxiliary program that is not implemented as a subroutine. Such assignments to real locations have no scientific significance, but add remarkably to the interest with which the outcome is received.) The footprint of the destructive blast wave from aerial explosions is calculated, using the customary nuclear weapons rule that overpressures of 4 psi (2.5×10^5 dyn cm^{-2} in the present simultations) suffice to demolish almost all normal (not hardened) buildings. Radiant heating of the surface beneath the entering body is summed to determine whether radiant ignition of surface materials occurs, again in the same manner as in nuclear weapons effects modeling. Impacts into the oceans are modeled both as hazards against local life and as generators of tsunami waves. The tsunami source (impact) location is assigned a distance to the nearest populated coastline by means of another random number that is fit to a function derived from global geographical data.

The total deaths and injuries are calculated by multiplying the areas affected by the local population density. Categories of deaths include direct blast effects causing building collapse, firestorms, tsunami inundation, and blast-wave-accelerated window glass fragments. Care is taken not to count deaths two or three times (those fatally injured by fire frequently are also fatally injured by glass, and those caught in collapsed buildings are almost always also within the area swept by firestorms).

At the end of each century run the individual outcomes are sorted to call out the three largest events in each of three categories: mass, energy, and (most useful for very large events) the iridium signature (subroutine SORT). Fatal events are flagged, and may be reported separately.

Output from the Monte Carlo simulation can be sent to screen, disk file, or directly to a printer. Formatting and management of data are accomplished by subroutine DATAPRINT.

The Typical Run

We will emphasize impact phenomena on natural human timescales. The minimum run is usually for $t = 100$ a (years) and $dt = 1$ a. The longest are usually for $t = 10^4$ a and $dt = 10^2$ a. Chapter 7 will report both the average results of running this program many times with different random number seeds and the dispersion of results experienced. Thus both the most probable outcome and the range of likely outcomes will be presented in such a way that the futility of making accurate predictions of the next century's impact fatality rate will be evident.

Upon beginning this effort, I speculated on what my reaction would (and ought to) be if, in one of my first thousand years of runs, I encountered a million-year energy event. Perhaps fortunately, no such highly improbable event has actually occurred. As in the real world, however, many kinds of comparably improbable events are possible, and many occur. Once three 1000-year events occurred in the same century, a coincidence that has the same probability (10^{-3}) as seeing one million-year impact in a 1000 years of runs. Sometimes very unusual sets of entry conditions, such as horizontal entry with skipout or capture into orbit, recurred in a short interval of time, or bodies of rare compositional types "ganged up" on Earth. But the overall statistics of all the runs conducted show no signs of unnatural bias by any statistical criterion I have checked. I have therefore compiled the results

Method of Calculation

given in Chapter 7 with confidence that they are free of identifiable bias and are statistically in accord with reality, and with my assurance that the results have not been edited or selected to make them either more or less uniform, diverse, interesting, boring, or cataclysmic.

7

Results

Results of 100 × 1-Year Runs

The results of 10 illustrative 100-year runs, with 1-year time steps, for a constant population of 5 billion people are presented in Tables 7.1 through 7.4. The 1000 events reported here were mostly airbursts. Each 100-year run is identified by a letter, and each event is identified by that letter and the year (1–100) in which it occurred. On the average, two fatal events occur per century. Because the events are stochastic, centuries without any fatal events are not unusual. (Lethal strikes on individuals by small subsonic meteorites, the only category of impact-related hazard neglected in this simulation, probably contribute about one fatality per century. This general conclusion is borne out both by preliminary numerical integrations and by the historical record.) The breakdown of these lethal events by type, given in Table 7.1, is instructive in that it suggests that a large preponderance (15 out of 20) of the fatal events on the century timescale are caused by airbursts. The population statistics of the lethal bodies do not resemble the statistics of the total incident flux for good reason: The largest bodies expected on the century timescale are in the size range of under 100 m, within which weak materials are common and atmospheric sorting of the incoming bodies is very important. The altitude distribution of these explosive events versus yield (total energy content) and compositional type is presented in Fig. 7.1.

A measure of the selectivity imposed by aerodynamic sorting is the prominence of strong impactors on the fatality list. Irons, which make up 3.2% of the meteorite falls, contribute 30% (6 of 20) lethal impactors. Achondrites, hard stones that make up 7.9% of the observed falls, contribute 25% of the lethal events, and stony-irons, making up 1.1% of the falls, contribute 10% of the fatal events. Thus the rare, strong meteorite types (a total of only 12% of the incident flux at the top of the atmosphere) account for 40% of the lethal events on a century timescale.

Although CI and CM carbonaceous chondrites make up more than 25% of the bombarding flux, on the century timescale they contribute no lethal events. Because of their low physical strengths, they are crushed by the aerodynamic ram pressure so high in the atmosphere that their blast waves, while noticeable and even spectacular at ground level, lack the intensity needed to demolish structures and inflict lethal injury. Note, however, that one fatal event (I23) is attributed to a small long-period comet fragment. Such a body would almost certainly be an extinct cometary core in the Apollo family, typically having an aphelion distance of about 5 AU. This body caused a 33-Mt explosion at 37 km of altitude that ignited a firestorm in a town. Weak materials that usually explode at high altitude, such as comet fragments and CI and CM carbonaceous chondrites, are proportionately less likely to impact the surface and more likely to initiate fires over large areas.

Comparing the frequency of fatal events in this simulation to existing historical records is not a trivial exercise. First, the population of Earth prior to the 19th century probably never exceeded 1 billion people. A very large proportion of these people were illiterate. According to the simulations, a significant proportion of past lethal events may have occurred under circumstances in which they were not recognized as impacts. If recorded at all, they may have been described as "tidal waves" or "earthquakes," or even, as in the *History of the Franks*, as "mysterious fire from Heaven." The geological record left by airbursts is extremely fragile, and tsunami events will generally bear no record of their cause. In addition, many lethal events may have killed their only eyewitnesses. Impact events that leave verifiable long-term evidence, such as craters and obviously nonterrestrial material samples, are not common on the century timescale. Nonetheless, when comparing these simulations to the late 19th and early 20th century, the number of fatalities and fatal events should be scaled down by about a factor of 3. For a comparison with the middle and early 19th century, a reduction of about a factor of 5 would be more appropriate. Over 1000-year spans, in which the simulation assumes a population of 5×10^9, the average population of Earth has always been less than 1×10^9. Fatality estimates from the simulation should therefore be reduced by a similar factor before comparing them to historical precedents.

We may see from Table 7.1 that only five lethal events in 1000 years of runs were attributed to bodies that were intact at

TABLE 7.1

Number of Fatal Events vs. Type of Impactor 1000 Years of Runs with Population 5 × 10^9

Event (run/year)	Type of impactor									
	SP comet	LP comet	CI	CM	OC	Ach	Iron	Pall	Mes	Sum
A/7										
B/29							1			
B/65						1*				
C/42							1			
C/63						1*	1			
D/46						1*				
D/54								1*		
E/22					1*					
E/79							1*			
F/30							1			

F/39						
F/93			1*			
G/10			1*			1
H/28		1*				
H/55		1*				
H/89		1*				
H/91		1*				
I/23	1*					
I/89		1*				
J/74				1*		
*Airbursts	1	6	5	2	1	15
Impacts	0	0	0	4	1	5
Sum	1	6	5	6	2	20

Key: SP = short-period comet, LP = long-period comet, CI = I-type carbonaceous chondrite, CM = M-type carbonaceous chondrite, OC = ordinary (H-, L-, or LL-type) chondrite, Ach = achondrite, Pall = pallasite, Mes = mesosiderite.

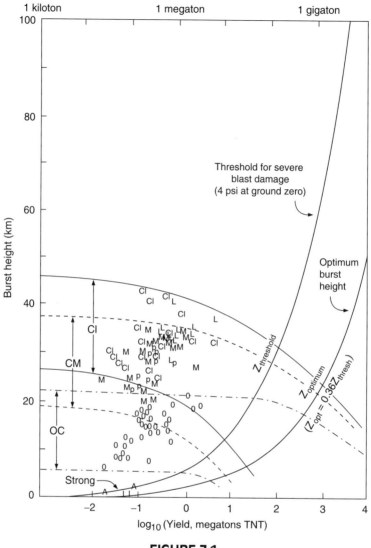

FIGURE 7.1

Results of a typical 100-year simulation run. The burst heights and yields for several classes of impactors are plotted on a graph with the same format as Fig. 4.1. The slowest bodies in each category penetrate deepest. Only the largest energy event of each year is plotted. The range of behavior seen over many runs is outlined. On the century timescale, the highest yield events are usually harmless high-altitude airbursts of CM chondrite, CI chondrite, or debris from long-period (L) and periodic (P) comets. As is typical in 100- and 1000-year runs, almost every life-threatening event is caused by rare, small bodies of exceptional strength (irons, achondrites, mesosiderites, pallasites).

the time of impact. Of these, four were irons and the fifth was a metal-matrix pallasitic stony-iron. Studies of chemical tracers in impact craters on Earth likewise show a striking enhancement of irons, achondrites, and stony-irons: Of 22 craters under 30 km in diameter that have been chemically examined, 6 impactors have been implicated as irons and 3 as achondrites (see Grieve and Shoemaker, 1994).

Another interesting result of these simulations is that the rare class of strong bodies can inflict severe, albeit local, surface damage even with rather small impact energies. Several prior studies have dismissed events with yields less than 1 Mt, or even 10 Mt, as ineffectual at causing fatalities. The ten 100-year runs reported here, however, include lethal events with total yields as low as 0.023 Mt, which corresponds very closely to the Hiroshima or Nagasaki atomic bomb. Interestingly, 10 of the 20 lethal events involve yields less than 0.1 Mt, and two others fall between 0.1 and 1 Mt. All 12 of these low-yield killers were rare, strong types: irons, stony-irons, or achondrites. All five lethal surface-impact events involved yields less than 0.054 Mt. By comparison, only 4 of these 20 lethal events are larger than 10 Mt (Table 7.2). The largest single lethal event in our 1000 years of runs (H28) was 103 Mt, which is about a 2000-year object. These Tunguska-type explosions bear striking similarity to the airburst model presented by Sekanina (1983).

The fatality rates generated by these events are summarized in Table 7.3. The number of fatalities in a single lethal event ranges from 1 to 237,376. The largest single event (in terms of lethality, not explosive yield) was responsible for 83% of the total number of deaths in the 1000 years of runs. The domination of the casualty figures by the one or two worst events is a general feature of all the simulations run to date.

Fatal events occurred in every one of these 10 runs. Runs A, G, and J had only a single fatal event in the century. Run H

TABLE 7.2

Number of Fatalities vs. Impactor Yield
Ten 100 × 1-Year Runs with Population 5 × 10^9

Event (run/year)	Impactor yield (Mt TNT)	Fatalities	Type
0.01–0.10 Mt			
C/42	0.023	628	Iron
F/93	0.023	810	Achondrite
A/74	0.025	1110	Iron
B/29	0.025	444	Achondrite
D/54	0.031	1	Pallasite
B/65	0.041	161	Iron
E/79	0.043	154	Iron
G/10	0.044	5019	Pallasite
F/30	0.054	1357	Iron
F/39	0.087	573	Achondrite
0.10–1.00 Mt			
C/63	0.144	19	Achondrite
J/74	0.369	87	Iron
1.00–10.0 Mt			
D/46	1.21	4146	Achondrite
H/55	2.24	5452	Ord. ch.
E/22	2.78	8591	Ord. ch.
I/89	3.68	1	Ord. ch.
10.0–100 Mt			
H/91	27.3	2388	Ord. ch.
I/23	33.2	2276	Long comet
H/89	82.6	237376	Ord. ch.
>100 Mt			
H/28	103.	12697	Ord. ch.

contained four astonishingly similar events that killed more than 2000 people each. All four were caused by low-speed entry of massive ordinary chondrite bodies over areas with moderate to high population densities. In two or three of these centuries, the evidence linking the deaths to impactors was so tenuous that the reality of the impact hazard could rationally be doubted. The fatal impact events in them could easily be attributed to "earthquake," "fire," or "storm surges."

The leading causes of death in these simulations are outlined in Table 7.4. The most dangerous killer is urban firestorms, which killed people in every run except A and G. Fully 80% of the fatalities were due to fire. Direct blast effects were second in importance, accounting for another 16% of the total. Tsunamis were third, at 2.6%, and blast-wave-accelerated window glass was fourth. The number exposed to glass shrapnel traveling at speeds of Mach 0.5 was actually much larger, but almost all of those severely injured by glass were in regions swept by firestorms. If fire were completely suppressed, the death toll due to glass would be greatly increased.

It is interesting that, even on this short a timescale, with the largest ocean impact weighing in at 0.054 Mt, tsunami waves caused three fatal events, one due to the chance fall of the body in busy fishing grounds, and another to a fall off the mouth of a river with densely populated banks. The significance of these tsunami events is not that they are the leading cause of death, but rather that they are a "stealthy" effect of impacts, in that the reported lethal events cannot as a rule be linked to an impact cause.

The average death rate seen in these runs was 283 people per year. In the 10 individual centuries, the fatality rate ranged from 0.9 to 2579 people per year. In the absence of the single most lethal event (a 1000-year event), the average fatality rate would be only 48 people per year. This wide range brings out clearly the enormous variability of the impact threat and the impossibility of predicting the fatality rate for any given century on purely statistical grounds: The variance is much larger than the mean.

TABLE 7.3

Number of Fatalities vs. Type of Impactor
Ten 100 × 1-Year Runs with Population 5 × 10^9

Event (run/year)	Type of impactor									
	SP comet	LP comet	CI	CM	OC	Ach	Iron	Pall	Mes	Sum
A/74							1110			
B/29						444*				
B/65							161			
C/42							628			
C/63						19*				
D/46						4146*				
D/54								1*		
E/22					8591*					
E/79							154*			

98

F/30						
F/39			1357			
F/93	573*					
G/10	810*			5019		
H/28		12697*				
H/55		5452*				
H/89		237376*				
H/91		2388*				
I/23	2276*					
I/89		1*				
J/74			87*			
*Airbursts	2276	266505	5992	241	1	275015
Impacts	0	0	0	3256	5019	8275
Sum	2276	266505	5992	3497	5020	283290

Key: SP = short-period comet, LP = long-period comet, CI = I-type carbonaceous chondrite, CM = M-type carbonaceous chondrite, OC = ordinary (H-, L-, or LL-type) chondrite, Ach = achondrite, Pall = pallasite, Mes = mesosiderite.

TABLE 7.4

Number of Fatalities vs. Cause of Death
Ten 100 × 1-Year Runs with Population 5 × 10^9

Event (run/year)	Cause of death				
	Blast	Fire	Tsunami	Glass	Sum
A/74			1110		1110
B/29	381	63			444
B/65	161				161
C/42	628				628
C/63		19			19
D/46		4146			4146
D/54		1			1
E/22		8591			8591
E/79	112	20		22	154
F/30	50		1307		1357
F/39	251	80		242	573
F/93	704	106			810
G/10			5019		5019
H/28	1431	10066		1200	12697
H/55		5452			5452
H/89	42386	194990			237376
H/91	3	2385			2388
I/23		2276			2276
I/89		1			1
J/74		87			87
Sum	46107	228283	7436	1464	283290

Figure 7.2 summarizes the mean fatality rates seen in 100 different 100-year runs. As expected, the largest one or two individual events dominate the overall statistics. The most probable outcome for one century is a mean death rate of about 20 people per year.

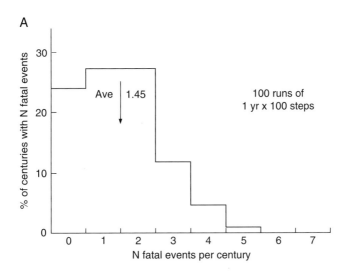

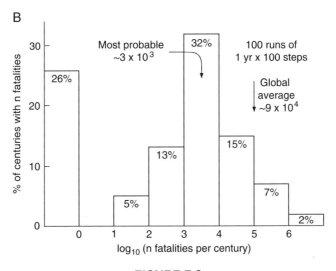

FIGURE 7.2

Histograms of fatality statistics for one hundred 100-year simulation runs. (a) Frequency distribution of the number of fatal events per century. (b) Frequency distribution of the mean fatality rate. The most deadly 10% of the centuries account for 90% of the fatalities.

Results of 1000 × 1-Year Runs

The results of ten 1000-year runs with 1-year time steps are summarized in Tables 7.5 through 7.13. The runs are numbered from 3000 to 3009, and each individual event is identified by the year of its occurrence in the 1000-year span of the run (such as 3002-447). In 10,000 years there should be an average of seven impacts larger than 100 Mt TNT and the largest expected impactor is about 900 Mt. In the Monte Carlo simulations, 10 explosions larger than 100 Mt were observed (of which only 4 resulted in fatalities), and the largest single explosion (number 3005-354) was 5610 Mt, which is about a 40,000-year event. The probability of such an event occurring in a 10,000-year run is therefore 0.25. The largest event was an ocean impact of a 4.67×10^{13} g ordinary chondrite, which began to break up at an altitude of 19.5 km. The second largest explosion was an ocean impact of an achondritic body.

There were 183 lethal impact events. The number of fatal events observed per century ranged from 0 to 6, with the most probable number being 2. The number of fatal events per millennium ranged from 13 to 24, with a mean of 18.3.

Of the various impactor types, 86.4% were asteroidal and 13.6% were cometary. The irons, which made up only 3% of the impactor flux, accounted for more than a quarter of all the fatal events. The breakdown of fatal events in all 10 runs according to impactor composition class is given in Table 7.5. Because of the great physical strength of most irons, even quite small irons can penetrate successfully to the ground, where they deliver a small explosive yield that has a devastating effect on a limited area. The four strongest classes, comprising the irons, pallasites, mesosiderites, and achondrites, account for 48.5% of the lethal

TABLE 7.5

Number of Fatal Events vs. Type of Impactor
Ten 1000 × 1-Year Runs with Population 5 × 10⁹

Run	\multicolumn{10}{c	}{Type of impactor}

Run	LP comet	SP comet	CI	CM	OCh	Ach	Mes	Pall	Iron	Sum
3000	3		1	1	9	1			6^{1a}	21
3001	1			2	5	3^2		4^1	6^2	21
3002	2	1			8	5			4	20
3003	1	2	2	1	5				3^2	14
3004	1				3	3	1		5^2	13
3005	2				5^2	4^1		2	2^1	15
3006	2	2			4	1			5^1	14
3007		1	1	1	7	5	2^1	1	6^3	24
3008	2	1		1	4	5		1	5	19
3009	2	2	1	1	7^2	2	1	1	5	22
Airbursts	16	9	5	7	53	26	3	8	35	162
Impacts					4	3	1	1	12	21
Sum	16	9	5	7	57	29	4	9	47	183

Key: SP = short-period comet, LP = long-period comet, CI = I-type carbonaceous chondrite, CM = M-type carbonaceous chondrite, OC = ordinary (H-, L-, or LL-type) chondrite, Ach = achondrite, Pall = pallasite, Mes = mesosiderite.

[a] Exponents are the number of surface impact events. Thus 5^2 should be read as five lethal events, two of which are surface impacts and the remainder are airbursts.

events even though they collectively account for only about 6% of the incident flux. The ordinary (H, L, and LL) chondrites accounted for another 31.1% of the lethal events. The very abundant CI and CM chondrites typically explode at high altitudes because of their low physical strength. Together they make up only 6.5% of the lethal events. There were never more than two C-type impactor events per millennium that resulted in fatalities, even though many of the explosions were very energetic.

Of the 183 lethal events, 21 were surface impacts and 162 were airbursts (Table 7.5). Of the 21 lethal surface impactors, 18 had impact yields of 133 kt or less, one delivered 29.4 Mt, and two were in excess of 1 Gt. Among the smallest impactors (impact yields less than 150 kt) the lethal surface-impact events were dominated by rare, strong materials: 12 of the 18 were irons, 2 were stony-irons (one mesosiderite and one pallasite), 2 were achondritic stones, and 2 were small ordinary chondrites. Of the three largest lethal impactors, two were ordinary chondrites and one was an achondrite. All of the CI, CM, and cometary fatalities were caused by airbursts, not surface impacts.

Although several irons larger than the 100-kt range also contributed to the fatality toll, these bodies exploded within 4 km of the ground and blasted the surface with a "shotgun pattern" of metallic shrapnel. Such events, though technically airbursts, can drop hypersonic fragments that are both large enough to excavate small craters and sufficiently distinctive in appearance to be recognizable as meteorites. These events are reminiscent of the Sikhote Alin iron meteorite fall near Vladivostok in 1947, in which craters up to 100 m in diameter were excavated, and many tons of iron meteorites were recovered.

It is interesting to analyze the circumstances that allowed the weakest of the impacting materials, the four ordinary chondrite impactors, to survive to the ground. The first, 3005-140, entered the atmosphere almost horizontally, at an entry angle of only

8.95 degrees, with an entry speed (at 140 km altitude) of 14.00 km s^{-1}. This 1650-tonne body was decelerated gently enough to avoid breakup during descent, and struck the surface at only 3.39 km s^{-1}, delivering a total impact energy of only 2.3 kt TNT, about that of a small tactical nuclear explosive. This event is reminiscent of the much larger ordinary chondritic body that struck the Argentine pampas almost horizontally several thousand years ago with a total yield of about 350 Mt (Schultz and Lianza, 1992). The second ordinary chondrite impactor (3005-354) was a giant body of 4.67 × 10^7 tonnes mass and an impact energy of 5610 Mt (5.61 Gt). This body entered at 31.72 km s^{-1} at an elevation angle of 27.58 degrees and began to disrupt at an altitude of 19.48 km, striking the ocean surface as a compact debris swarm with most of its initial energy content intact. The entering asteroid had a diameter of 286 m.

The third ordinary chondrite impactor, 3009-021, massed 1910 tonnes at entry, and entered at an angle of only 9.08 degrees below the horizontal. It, like 3005-140, decelerated without disruption from its original entry speed of 14.99 km s^{-1} to a final speed of 3.41 km s^{-1}, striking land intact with an impact energy of 2.7 kt. The final example, 3009-203, a 1.78-million-tonne asteroid, entered at a very typical angle (47.01 degrees) but at nearly the lowest possible speed, only 11.77 km s^{-1}. It began to disrupt at an altitude of 9.9 km, and struck the ocean surface as a compact debris swarm with a kinetic energy of 29.4 Mt.

It is noteworthy that all four of these successful surface impactors were statistical outriders of the population: 3005-140 and 3009-021 entered at exceptionally low entry angles, allowing them ample opportunity to decelerate smoothly. The largest, 3005-354, was of extraordinary size, a 40,000-year object nearly 300 m in diameter, and the fourth, 3009-203, entered at an extremely low speed, little more than escape velocity. Such a low entry speed is characteristic of the small low-eccentricity near-

Earth asteroids found by Spacewatch, tentatively called *Arjunas* by Gehrels. In calculations of "typical" ordinary chondrite entry behavior, using average entry velocities, 45° entry angles, and small sizes appropriate to, say, 1000-year objects, none of these four lethal events that occur in a realistic Monte Carlo simulation would have been recognized. The casualty toll from ordinary chondrite bodies would have been estimated at 0 instead of a toll of 11,287,629 found by the detailed simulation.

The impactor classes ranked in decreasing order of lethality place the abundant and moderately strong ordinary chondrites first in effect, followed by the long-period comets with their rare, devastatingly violent high-altitude detonations and firestorm ignition. All the other classes of impactors combined (in descending order, the achondrites, CI chondrites, irons, short-period comets, CM chondrites, then the rare mesosiderites and pallasites) contribute about as many fatalities as the long-period comets. But these statistics are dominated in each composition class by one single event that accounts for a disproportionate share of the fatalities. The sum of the most lethal single events in each category accounts for fully 86.5% of all the fatalities found.

The distribution of explosive yields among the fatal events was very broad. Table 7.6, which displays the results for the number of fatal events versus impactor yield and run number, shows that 2 lethal events (the ordinary chondrites 3005-140 and 3009-021 discussed earlier) delivered less than 10 kt of explosive yield, 44 delivered between 10 and 100 kt, 32 between 100 kt and 1 Mt, 58 between 1 and 10 Mt, 41 between 10 and 100 Mt, 2 between 100 Mt and 1 Gt, and 2 between 1 and 10 Gt. The breakdown of fatality rate versus impactor energy and compositional type is detailed in Table 7.7. The two smallest yields correspond to severely decelerated ordinary chondrite bodies that entered nearly horizontally. Clearly the lethal events with 1-Mt

TABLE 7.6

Number of Fatal Events vs. Impactor Yield and Run Ten 1000 × 1-Year Runs with Population 5 × 10⁹

	Yield range (Mt TNT)						
Run	10^{-3}–10^{-2}	10^{-2}–10^{-1}	10^{-1}–1	1–10^1	10^1–10^2	10^2–10^3	>10^3
3000		5	3	6	7		
3001		10	1	5	3		1
3002		3	5	7	5		
3003		3		6	5		
3004		4	3	5	1		
3005	1	3	5	2	3		1
3006		2	4	4	3	1	
3007		7	5	7	5		
3008		2	5	7	5		
3009	1	6	1	5	8	1	
Sum	2	45	32	55	45	2	2

yields are dominated by rare, intrinsically strong objects. Note the great importance of cometary events in the 10- to 100-Mt range: Smaller cometary bodies explode at high altitudes with too little energy to affect the ground, whereas larger bodies (>100 Mt) are relatively rare, and did not occur in the 10^4 years of simulation. The more competent ordinary chondrite bodies penetrate deeper before exploding, and therefore can have lethal effects down to several hundred kilotons of yield. Note that the very small yield ordinary chondrite objects (<10 kt) have lethal effects for wholly different reasons than their larger cousins: rare, very shallow entry angles with deceleration and deep penetration for the smallest, and moderate-altitude airbursts for the megaton-class bodies.

TABLE 7.7
Number of Fatal Events vs. Impactor Yield and Type
Ten 1000 × 1-Year Runs with Population 5 × 10⁹

Type	Yield range (Mt TNT)						
	10^{-3}–10^{-2}	10^{-2}–10^{-1}	10^{-1}–1	1–10^1	10^1–10^2	10^2–10^3	>10^3
LP comet	0	0	0	0	16	0	0
SP comet	0	0	0	2	6	1	0
CI	0	0	0	2	3	0	0
CM	0	0	0	3	3	1	0
OCh	2	0	4	36	14	0	1
Ach	0	10	9	8	1	0	1
Mes	0	4	0	0	0	0	0
Pall	0	6	3	0	0	0	0
Iron	0	25	16	4	2	0	0
Sum	2	45	32	55	45	2	2

Key: SP = short-period comet, LP = long-period comet, CI = I-type carbonaceous chondrite, CM = M-type carbonaceous chondrite, OC = ordinary (H-, L-, or LL-type) chondrite, Ach = achondrite, Pall = pallasite, Mes = mesosiderite.

The distribution of fatalities over impactor explosive yield and type is given in Table 7.8. The lethal effects of small (<1-Mt), strong impactors account for only 1% of the total fatalities, even though they involved nearly half the events listed in Table 7.7. The reason for the small death toll from these bodies is simply that they devastate relatively small areas for each event, much in the manner of the Sikhote Alin iron meteorite fall of 1947. The large majority of the deaths are due to one ordinary chondrite body of exceptional size and to a barrage of airbursts in the 10- to 100-Mt class. These airbursts are in the intermediate range of size in which aerodynamic disruption no longer affords full protection against fragile impactors, but in which the bodies are not yet so massive that they penetrate freely to the ground irrespective of physical strength.

The statistical variability of the fatal event rate is demonstrated by Table 7.9, which gives the number of fatal events for each century of the 10 millennial runs. The number of lethal events per century was found to vary from 0 to 6. The statistics of these event numbers are summarized at the bottom of the table: The most probable number of lethal events per century is 2 (average 1.83). Fully 43% of all the centuries in the simulation had 0 or 1 fatal event.

The detailed breakdown of number of fatalities versus the impactor compositional type and run number is given in Table 7.10. The millennial total death rates vary over a factor of 500.

The 10 most lethal events encountered in the simulation are summarized in Table 7.11. The most lethal single event, 3005-354, killed more than 11 million people, and the second worst, 3000-466, killed 830,000. Ten events killed more than 100,000 each. Of these 10 worst disasters, 8 (all except 3005-354 and 3001-841) were airbursts over land. The proximate cause of death for these events was firestorms ignited by radiation from

TABLE 7.8

Number of Fatalities vs. Impactor Yield and Type
Ten 1000 × 1-Year Runs with Population 5 × 10^9

Type	Yield range (Mt TNT)						
	10^{-3}–10^{-2}	10^{-2}–10^{-1}	10^{-1}–1	1–10^1	10^1–10^2	10^2–10^3	>10^3
LP comet	0	0	0	0	1081494	0	0
SP comet	0	0	0	5756	47917	145379	0
CI	0	0	0	2941	221895	0	0
CM	0	0	0	8158	25870	116327	0
OCh	370	0	1021	127812	1081760	0	11261000
Ach	0	12295	21398	17033	2169	0	272220
Mes	0	12366	0	0	0	0	0
Pall	0	9893	538	0	0	0	0
Iron	0	25468	36909	17225	120732	0	0
Sum	370	59395	59866	295242	2588837	261706	11533220

Key: SP = short-period comet, LP = long-period comet, CI = I-type carbonaceous chondrite, CM = M-type carbonaceous chondrite, OC = ordinary (H-, L-, or LL-type) chondrite, Ach = achondrite, Pall = pallasite, Mes = mesosiderite.

TABLE 7.9

Number of Fatal Events per Century
Ten 1000 × 1-Year Runs with Population 5×10^9

Run	Century										Sum
	1	2	3	4	5	6	7	8	9	10	
3000	3	2	5	0	3	2	2	3	1	0	21
3001	2	2	2	4	1	0	2	4	2	2	21
3002	4	2	0	2	2	2	5	1	1	1	20
3003	4	4	0	0	0	2	1	1	2	0	14
3004	2	2	1	3	0	2	0	2	1	0	13
3005	1	1	0	3	1	2	3	1	3	0	15
3006	0	2	1	2	2	0	1	4	1	1	14
3007	6	1	2	4	1	3	2	2	1	2	24
3008	0	1	4	3	3	1	3	1	1	2	19
3009	5	2	4	2	3	0	1	2	2	1	22
Sum	27	19	19	23	16	14	20	21	15	9	183

Number of fatal events per century: 0 1 2 3 4 5 6 7+
Number of occurrences: 17 26 32 12 9 3 1 0

TABLE 7.10

Number of Fatalities vs. Type of Impactor Ten 1000 × 1-Year Runs with Population 5 × 10^9

Run	SP comet	LP comet	CI	CM	OC	Ach	Iron	Pall	Mes	Sum
3000		834227	167920	20186	246813	1	5971			1275118
3001		879	5784	17909	272651	15351	9624			322198
3002	11264	116270			74595	28645	1156			231930
3003	10933	36740	19688	1458	32763		419			93001
3004		34			17008	4624	1456		1621	24743
3005		10015			11262414	8488	8428	27		11289372
3006	151024	71667			139896	16068	25274			403929
3007	1		1177	6472	702104	4734	1037	45	113	715689
3008	19094	4472		228	10850	8454	122172	51		165321
3009	6740	7178	45071	116327	48495	6406	18345	484	10632	259678
Sum	199056	1081482	224856	150455	12552847	350071	199609	10231	12366	147800973

Key: SP = short-period comet, LP = long-period comet, CI = I-type carbonaceous chondrite, CM = M-type carbonaceous chondrite, OC = ordinary (H-, L-, or LL-type) chondrite, Ach = achondrite, Pall = pallasite, Mes = mesosiderite.

TABLE 7.11
Ten Most Lethal Events
Ten 1000 × 1-Year Runs with Population 5 × 10^9

Rank	Event number	Type	Yield (MtTNT)	Log S	Fatalities	Fate	Principal killer
1	3005-354	OCh	5610.	2.445	11,261,000	oi	Tsunami
2	3000-466	LP	42.6	3.109	829,790	lf	Fire
3	3007-088	OCh	62.6	2.680	694,444	lf	Fire
4	3001-841	Ach	1830.	0.775	272,220	oi	Tsunami
5	3000-123	OCh	22.9	2.636	218,609	lf	Fire
6	3000-771	CI	18.3	2.764	167,920	lf	Fire
7	3006-951	SP	124.	1.752	145,379	lf	Fire
8	3006-744	Ch	10.4	2.947	138,660	lf	Fire
9	3009-707	CM	126.	1.590	116,327	lf	Fire
10	3002-423	LP	38.6	2.324	116,240	lf	Fire

Key: Fates are l = land, o = ocean, f = fragmented (airburst), i = impacted.

the airburst. The other two, which happened to be the two with the largest yield, were ocean impacts that struck near inhabited coastlines and drove tsunami waves ashore. Interestingly, 1 of the 10 most lethal events (the ordinary chondrite body 3006-744) had an explosive yield of a paltry 10.4 Mt. The reason for the high fatality rate is that the population density at ground zero for this event was 884 people per square kilometer. The modestly larger cometary airburst event 3000-466 achieved its status as the second most lethal event by striking an urban area with a population density of 1285 km^{-2}.

Four of the 10 most lethal events were caused by ordinary chondrite airbursts ("Tunguskas"), 3 by comets (2 long-period and 1 periodic), and 1 each by CI and CM chondrites and an achondrite. The broad representation of compositional types in this list is due to the domination of the death toll in the 10^4 year simulation by 100-Mt-class impactors. As noted earlier, these bodies are large enough to have lethal effects even if they detonate at high altitude.

There is no simple proportional relationship between fatalities and explosive yield. It is true that events of sufficient violence (1 Gt; roughly 250-m entering bodies) are invariably highly lethal no matter where they land. But at sizes below about 100 m (roughly 150 Mt) the lethality is a convolution of the population density and the strength of the impactor. The two most violent explosions in the simulation, 3005-354 and 3001-841, were the first and fourth most lethal events. The two 100- to 1000-Mt explosions, 3006-951 (a short-period comet 124-Mt airburst at 29.7 km) and 3009-707 (a CM chondrite 126-Mt airburst at 23.8 km) were only the seventh and ninth most lethal events. The second and third most lethal events were airbursts in the 50-Mt class, but both inflicted firestorms on densely populated areas. In many locations on Earth, such events would cause no fatalities.

In examining the historical record to identify such events during the past 10,000 years of Earth history, we are limited by a variety of factors. First and most obvious is the low population density and low level of literacy over most of Earth's surface for most of this period. It is also instructive to note that 8 of the 10 most devastating events were airbursts that would have left no obvious meteoritic material on the surface. The other 2 surface impacts struck in the deep ocean, where neither cratering nor survival of a recognizable sample is likely.

The total fatality rate during the 10,000 years of simulation was 1479 per year, with 76.5% of the fatalities due to the most deadly single impact. Without that event, the average death rate was 348 per year. Table 7.12 demonstrates that, for each composition class, the largest single object tends to dominate the death rate. The sole apparent exception, iron meteorites, is due to a

TABLE 7.12

**Predominance of Largest Event vs. Type of Impactor
Ten 1000 × 1-Year Runs with Population 5×10^9**

Type	Total fatalities	Most lethal event	% of total
LP comet	1,081,482	829,720	76.7
SP comet	199,056	145,379	73.0
CI	224,856	167,920	74.7
CM	150,455	116,327	77.3
OCh	12,552,847	11,261,000	89.7
Ach	350,071	272,220	77.8
Mes	12,366	10,632	86.0
Pall	10,231	6,908	67.5
Iron	199,609	89,579	44.9

Key: SP = short-period comet, LP = long-period comet, CI = I-type carbonaceous chondrite, CM = M-type carbonaceous chondrite, OC = ordinary (H-, L-, or LL-type) chondrite, Ach = achondrite, Pall = pallasite, Mes = mesosiderite.

perfectly plausible chance near equality of the two most severe events.

Finally, Table 7.13 examines the worst event in each millennium for the purpose of illustrating the variability of outcomes. The worst lethality in a millennium ranges from 11,000 to 11 million, covering a range of a factor of 1000. Note that run 3005 would have been completely average except for event 3005-354. The most sobering note is that 1 millennium (run 3004) experienced its most lethal event from an explosion with a yield of only 2.1 Mt. Five other millennia of the 10 reported here experienced their worst death toll from explosions smaller than 100 Mt. Therefore averting the worst disaster of each millennium would in the majority of cases require finding, predicting, and diverting bodies smaller than 100 m in diameter. Finding such small bodies presents significant difficulties.

Results of Longer Runs

It is clear that rare, very energetic events must dominate the fatality statistics even more strongly for timescales of 10^5, 10^6 years, or longer: 10^5-year impactors are so massive that they reliably penetrate to the surface and deliver their kinetic energy to the biosphere, irrespective of their composition, orbit, or entry conditions. To explore this matter, a number of runs have been conducted with extended timescales, running in some cases up to 10^6 years with time steps that are usually 10^{-2} of the length of the time interval surveyed (that is, only the largest event in each 10^4-year time step is reported in a 10^6-year run). The results of these runs generate considerable data on the way in which the average annual death rate increases when large time intervals are considered, and rare, very energetic events occur. These simula-

TABLE 7.13
Worst Fatal Event in Each Run
Ten 1000 × 1-Year Runs with Population 5 × 10^9

Run-event	Type	Fate	Yield (Mt TNT)	Fatalities in event	Total fatalities for run	% of total fatalities
3000-466	LP	lf	42.6	829,790	1,275,118	65.1
3001-841	Ach	oi	1830.	272,220	322,198	84.4
3002-423	LP	lf	38.6	116,240	231,930	50.1
3003-113	LP	lf	60.6	36,740	93,001	39.5
3004-881	OCh	lf	2.1	11,755	24,743	47.5
3005-354	OCh	oi	5610.	11,261,000	11,289,37	99.7
3006-951	SP	lf	24.	145,379	403,929	36.0
3007-088	OCh	lf	62.6	694,414	715,689	97.0
3008-368	Iron	lf	16.5	89,579	165,321	54.2
3009-707	CM	lf	126.	116,327	259,678	44.8

Key:
Types are LP = long-period comet, Ach = achondrite, SP = short-period (periodic) comet, CM = M-type carbonaceous chondrite, OCh = ordinary (H, L, or LL type) chondrite.
Fates are l = land, o = ocean, f = fragmented (airburst), i = impacted.

tions are summarized in Figs. 7.3 and 7.4. Figure 7.3 shows how the cause of death depends on the time interval covered by the simulation. On the century timescale, airbursts over continental population centers and "iron rain" from low-altitude airbursts are the dominant killers. Over longer periods of time, in which gigaton ocean impacts become important, the principal cause of death becomes tsunamis. At very long run times the main cause of death becomes climate catastrophe induced by global killers larger than about a million megatons. The linkage of fatalities to particular composition classes of impactors is explored in Fig. 7.4 as a function of the run time. Note the extreme importance of rare, strong bodies in the short-term death statistics. Over

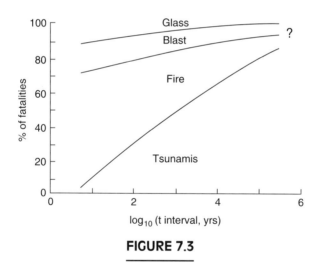

FIGURE 7.3

Cause of death statistics for runs of various lengths. Note the importance of direct local effects (glass acceleration, blast-driven structural failure, and firestorm ignition) over short timescales and the increasing dominance of tsunamis at longer time scales. Beyond 10^5 years, at a threshold that remains poorly understood (and is probably not very sharply defined), global climate crashes due to impact-injected dust or sulfate aerosols that cause mass deaths by famine.

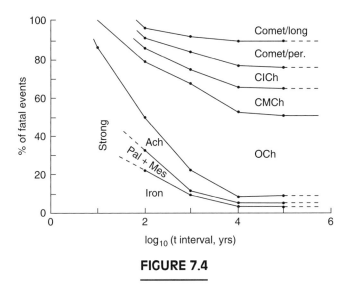

FIGURE 7.4

Types of bodies that cause fatal events. The disproportionate importance of strong bodies on all timescales of less than 10^4 years is clear. Fatal events on the 10-year timescale are so rare that the statistics are poor: irons, stony-irons, and achondrites have been lumped together as "strong" bodies in that time bin.

longer periods of time the largest impactors become so large that atmospheric sorting is no longer possible. The statistical distribution of lethal impactor types then becomes identical to the statistics in the bombarding population (which, we should recall, vary with size).

The average death rate over very long time periods, including global-scale killers, is several thousand people per year.

Unlike previous studies, the present simulation finds no sharp increase in fatality rates at the onset of global-scale effects. The reason is that those studies were based on highly oversimplified models of the impacting population, their entry conditions, and the target. The Monte Carlo simulations presented here, taking the natural variability of the population and the entry conditions fully into account, show convincingly that, on century or millennium timescales, most fatalities are caused by statistical outriders, not bodies with "average" physical and orbital traits. It should be obvious that casualty estimates based on average properties must be very incomplete. These statistically averaged calculations must, however, agree with the present simulations over very long run times. The reason is that, over tens to hundreds of thousands of years, most of the casualties are due to very large events, for which selection effects are negligible. We have already seen that the very largest events in a 10,000-year interval are so large that atmospheric sorting is ineffectual—the largest bodies in even longer runs are so large that they must penetrate deeply. Over shorter time intervals, however, the largest expected bodies are still so small that atmospheric sorting is very important: "Average" bodies do not penetrate, whereas the strongest bodies in a realistic distribution do penetrate successfully. Thus the difference between a statistically averaged population model and a detailed population distribution model should disappear for long run times. Clearly, however, the Monte Carlo treatment must predict more casualties over shorter time intervals.

With these distinctions in mind, the results of these extended calculations may be compared with the predictions of published statistical models of the impact hazard. Not surprisingly, long-term fatality rate averages from the simulation runs reported here are indeed very similar to those projected by purely statistical models (Chapman, 1993; Chapman and Morrison, 1994; Morrison *et al.*, 1994). The fatality rate as a function of the duration of the time interval found in these simulations is compared to the statistical projections of Chapman and Morrison in Fig. 7.5. There are, as expected, large and interesting differences at shorter timescales, where the Monte Carlo model predicts higher fatality

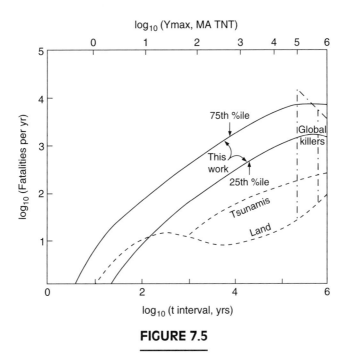

FIGURE 7.5

The dependence of average annual fatality rate on duration of exposure to the bombarding flux. The results of the present simulation runs are compared to the statistical predictions of Chapman and Morrison, as cited in the text. Note that arguments based on average impactor properties (velocity, strength, entry angle) are accurate for long time periods, but fail to reflect the complexity of short-term events. (From Morrison et al., 1994.)

rates. These predictions follow from the explicit inclusion in the present simulation of rare types of impactors and the complexities inherent in realistic distributions of compositional types, entry velocities, entry angles, impactor location, and fragmentation behavior.

8

Implications for Hazard Assessment and Abatement

Who knows whether, when a comet shall approach this globe to destroy it, as it often has been and will be destroyed, men will not tear rocks from their foundations by means of steam, and hurl mountains, as the giants are said to have done, against the flaming mass?—And then we shall have traditions of Titans again, and of wars with Heaven.

Lord Byron, 1821

Assessment of the Impact Hazard

Several steps are necessary to identify impact hazards. First, and most obvious, is the necessity of discovering bodies on

Earth-crossing trajectories. This problem has been studied by several workshops, and the technology requirements for this effort are reasonably well understood.

The second step is to determine the orbit of the body with sufficient accuracy so that it can be reacquired at the time of its next close passage through Earth's nighttime skies. The ultimate goal is to refine the orbit estimation so that future, very close approaches to Earth can be predicted well enough to identify possible opportunities for collision. Note that it is not computationally practical to calculate the impact of a body a millennium in advance: A body capable of impact is likely to make many close approaches before impacting, and therefore must exhibit chaotic behavior. Nonetheless, the next approach close enough to justify increased scrutiny can readily be predicted centuries in advance.

Third is the physicochemical characterization of the body to understand its probable response to atmospheric entry. Curiously, this sort of information is of least importance for the largest potential impactors, which are relatively easy to observe, and of greatest importance for bodies smaller than about 100 m, for which obtaining such data is very difficult. The most elementary characterization involves estimating the size and mass of the body and measuring its light curve, which tells the rotation period. Ideally, further characterization should involve studies of the reflection spectrum to identify minerals and, with luck, to link the body with a known class of meteorites whose detailed chemical and physical properties can be studied in the laboratory. In the event of a close flyby, Earth-based radar may be used to further characterize the shape and electrical properties of the surface. Radar has already been used to demonstrate that at least one near-Earth asteroid has a metallic surface (Ostro *et al.*, 1991).

If an Earth-crossing near-Earth object (NEO) is found to be following an orbit that will take it dangerously close to Earth, it would be enormously useful to dispatch a small dedicated spacecraft to that asteroid to carry out several crucial tasks. Such a probe could provide detailed chemical and physical data on the body, including its mineralogy, shape, density, rotation, gravity field, presence or absence of regolith, and so on. Precise future tracking of the NEO would be made possible by depositing a radar transponder on its surface. Because sample return from many NEOs is relatively easy, it may also be desirable to land, sample the surface, and return a small sample (less than a kilogram) to Earth.

In some circumstances, especially when a very close encounter (a potential impact) is expected, it may be desirable to carry out a deflection experiment on another, nonthreatening NEO before attempting to deflect the threatening body. Ideally, this experiment should be conducted on a body whose orbit rules it out as a foreseeable impact hazard. Qualifying traits would be high orbital inclination and wide separation of both of the orbital nodes from 1.000 AU.

Limitations on Detection Strategies

Thorough cataloging of the Earth-threatening population on a timescale of 10 years is entirely feasible, with the exception of certain minor categories of object. Perhaps the greatest single shortcoming of existing schemes for hazard assessment is in the inability of present search techniques to detect long-period bodies approaching the day side of Earth. Bodies in periodic, Earth-crossing orbits are far more likely to miss than strike Earth, and will from time to time pass through the nighttime sky, where a diligent search program will see and report them. But long-period comets, those with orbital periods of 10,000 to millions of years, have never passed Earth before in historical times. After once flying by Earth, they will generally not be back for another several million years. Further, their orbits are subject to large disturbances by nongravitational forces and by planetary gravitational perturbations, making any attempt at accurate long-term prediction of their next return (millions of years from now) both practically useless and mathematically futile. Fortunately, the probability of a large long-period Earth-crossing comet striking Earth on a given perihelion passage is extremely low, on the order of 10^{-9}. Given 10 perihelion passages of kilometer-sized long-period comets per year, the mean time between major impacts is 10^8 years. In addition, at least 90% of all approaching long-period comets should approach from other directions, and hence be readily discovered months before node crossing. The probability that a global-scale lethal event will be caused in the next year by an unseen large comet is again about 10^{-9}.

The most intimidating of the threatening bodies that would remain undiscovered after a 10-year intensive search campaign are small long-period comets, which account for 1.6 lethal events per millennium in the simulations. These bodies, which are typically on the order of 100 m in diameter or less, are roughly 10,000 times as common as the kilometer-sized global killers. In other words, the mean time between two consecutive lethal impacts of long-period comets is about 600 years. The probability that a cometary impact with mass mortality has occurred in human history is therefore of order unity. Of the 16 such events encountered in 10,000 years of runs, 2 killed more than 100,000 people and 2 others killed more than 10,000. All 16 of these are fairly characterized as local, not regional or global, threats. The three least disastrous lethal comet-impact events together killed a total of less than 100 people. In only 2 of these 10 millennia did long-period comets account for either the majority of deaths or the largest single lethal event. The largest two events together accounted for only 6.4% of all casualties, and all the long-period comet events combined contributed only 7.3% of the death toll. Of these bodies, warning times could range from 0 to about 2 years, with about 1 year being typical. This short a warning time imposes severe restrictions on defensive activities, but in almost every case there will be sufficient warning for civil defense purposes, principally evacuation of the areas most at risk. Cometary impacts without warning probably account for less than 10% of the comet impact rate, suggesting a residual death rate due to small long-period comets not detected before impact that is on the order of 0.6–0.7% of the total death rate from all impactors of all types over a 10,000-year time span. Over million-year or longer timescales the largest impactor is expected to be a comet. The probability that this event will occur without warning is a few percent. Therefore the problems of detection and prediction of long-period comets becomes serious only over very long time spans of millions of years.

There are so many other much more immediate types of impact hazard that fixation on this small part of the problem is completely unjustified. Likewise, building a detection or interception strategy to meet such a rare threat makes no sense until the far more pressing, far easier task of dealing with periodic (NEA and short-period comet) Earth-crossers is accomplished.

A second category of impactors is also capable of impact without prior warning: NEOs in pathological orbits. Several possibilities come to mind. First, there are Arjuna-type asteroids with synodic periods longer than 10 years. These bodies may never pass close enough to Earth for discovery during a decades-long search campaign, and hence have a greater likelihood of remaining undiscovered before impact. In addition to the Arjunas themselves, the presence of some asteroids in 1 : 1 resonant orbits is certain, both in the L4 and L5 points on Earth's orbit and in "horseshoe" orbits spanning L4 and L5. I refer here to the recent discovery that the asteroid 3753 1986TO is in a horseshoe orbit (Wiegert *et al.*, 1997).

Second are asteroids of as-yet unverified orbital classes that barely graze Earth's orbit at their aphelion, and hence never appear in the night sky. They may, for example, be in Aten-like orbits with their aphelia roughly aligned with Earth's aphelion at present. Only after prolonged periods of differential orbital precession may the lines of apsides of these bodies drift far enough to permit them to intersect Earth's orbit, at which point they become both detectable in the night sky and hazardous to Earth. A search program limited to the antisolar point, which depends on the opposition effect to maximize the brightness and detectability of its targets, would be incapable of detecting such bodies. Because we currently have no way to detect them, and have never seen one, we have no statistical data on their abundance.

The third category consists of bodies in conventional Apollo or Aten orbits that are so small that they must come very close to Earth to be detectable. These bodies, all less than about 100 m in size and frequently smaller than 10 m in diameter, are strictly local threats, with explosive yields ranging from about 100 Mt down to kilotons. They are so numerous and so hard to detect that the cost of finding and tracking them may be prohibitive. Together, they account for 179 of the 183 fatal events found in the 10,000-year simulation, and for 20.2% of the total fatalities. Despite these statistics, none of these bodies represents a global threat to human civilization. It is sobering to note that the most lethal incident in run 3006 was a mere 10.4 Mt, smaller than the Tunguska object of 1908. By ill fortune, 3006-744 happened to strike a target with a population density of nearly 1000 persons per square kilometer, killing 138,000 people. Such a small object could easily avoid detection and prediction.

Another diverting but irrelevant topic concerns Apollo and Aten objects that approach Earth on the day side, outbound from the Sun. It is true enough that such a body may be on an impact trajectory and may approach Earth invisibly. But such bodies also pass aphelion countless times before a collision occurs. Any systematic search effort would discover such a body far in advance of the threatened impact, and we would know where it was and where it was headed long before the threatened impact. Taking as a worst case an Arjuna with an expected lifetime of 10^5 years (most Apollos have lifetimes of order 3×10^7 years), the probability that we would discover it less than 100 years before impact is 10^{-3}. Of course, if we do not search, the probability of an unannounced impact is near 99.9%, not 0.1%.

It is imperative that search and tracking schemes be predicated on the principle of cataloging and understanding the larger members of the near-Earth population well in advance of any impact threat. The idea that we should, or must, hunker down and watch incoming bullets, detecting them for the first time just days or hours before impact, is an absurdity. All the NEAs capable of causing devastating global effects (those more than about 1–2 km in diameter) are readily detectable and, once discovered, easily tracked. They constitute the heart of the threat. But we already have the technology in hand to find, track, and predict them. Bodies in the 250-m size range, being far fainter and far more numerous, represent a much more serious, but still trackable, challenge. If the NEA population can be cataloged down to 100 m in diameter, then the expected death toll over 10,000 years would be reduced by about 79%. Further, and even more important, such an effort would provide a virtually complete census of the large (>1-km) asteroids capable of global extinction effects.

It is also important that we recognize the great improbability of an undiscovered global-killer long-period comet impact within the next 10 millennia ($P < 10^{-4}$). The vast majority of the threatening population of global killers can be efficiently detected and predicted using known technologies. It would be folly to fail to act in our own defense simply because our protection from global killers would be "only 99.9% complete." Assurances of absolute certainty can be offered only by charlatans and deities.

Economics of Search and Tracking

How sensitive a search effort should we mount? We could, for example, concentrate on the discovery and tracking of only those bodies large enough to be global killers. Or we could extend the search effort to smaller, fainter, more numerous bodies that are capable of regional tsunami threats. In the most ambitious scenario, we could extend the search to the enormous population of very smallest bodies that are capable of devastating a city or town. These three scenarios have very different implications for the equipment required, the scale of the search effort, and its cost of execution. For convenience in the following discussion, I define each of these categories in terms of a particular size threshold.

GLOBAL KILLERS

The threshold for global killers is usually taken to be 1 to 2 km in diameter. The definition is a bit arbitrary, because the lethality of the impact depends on where and when it lands, and to a smaller degree on its composition (density). There are conceivable circumstances in which a slightly smaller body might be equally damaging. Nonetheless, for the sake of concreteness, we shall take 1 km as the threshold size for devastating global effects. The population of NEOs is fairly well known from the various sources of data discussed in Chapter 2: There are about 2000 kilometer-sized NEAs and extinct short-period comets in Earth-crossing orbits. The mean lifetime of these bodies is roughly 50 million years, of which about a quarter will collide with Earth. Thus some 500 large NEOs strike Earth in a 50-million-year period, for a mean interval of about 100,000 years between consecutive impacts. The average impact energy of these kilometer-sized bodies is about 10^5 Mt per event. The probability that such an event will occur in your lifetime is approximately 0.1%. There is a 1% chance of such an event happening in any given millennium.

At present, approximately 8% of the population of kilometer-sized NEAs has already been discovered. The astronomical community has carefully examined the technical requirements for discovery of the remainder of these bodies and has concluded that present technology could accomplish a 90% census of these bodies in a decade (roughly 99% in two decades) at a cost of $3.5 to $5.0 million per year.

The average consequences of a 100-Gt explosion are fairly well understood. The death toll would be approximately 10^9 people, distributed rather evenly over the entire globe, and the economic loss would be about 10^{14}. Events of this scale threaten human civilization. The annualized cost of such impacts is 10,000 lives and 10^9 per year. The cost of discovering and tracking this population, $5 million per year, is 200 times smaller than the expected benefits that would accrue from averting a single impact.

The total cost of the survey ($5 million per year over 20 years) is one-millionth of the expected benefits during the next 100,000 years.

REGIONAL HAZARDS

We have seen from the simulations that bodies with dimensions of about 100 m are capable of generating devastating tsunamis that sweep the rim of an entire ocean basin. The average impact energy of bodies of this size is about 100 Mt. The total number of bodies with sizes between 100 m and 1 km is about 10^5, and the mean interval between 100-Mt impacts is therefore about 2000 years.

Although a body of this size could not cause global devastation, it could kill as many as 10^7 people and cause property damage of 10^{12}. The annualized loss due to regional impacts is about $500 million per year. By their nature, the consequences of regional impacts are distributed very nonuniformly about the globe. It is tempting to argue, for example, that the United States covers only 2% of the surface area of Earth, and therefore would almost certainly be spared the full weight of the impact. This attitude regards the threat to others as "tough luck" or "none of our concern." Quite aside from the moral bankruptcy of such a position, it is logically untenable. Fully 72% of all impacts strike the ocean surface, and roughly half will strike either the Pacific or Atlantic Ocean. At risk from ocean impacts are all assets close to sea level around the entire periphery of the ocean basin in which the impact occurs. The United States is, by this measure, one of the most vulnerable nations on Earth, since it has numerous major cities close to sea level on two separate oceans. The principal cities at risk are Anchorage, Honolulu, Hilo, Seattle, Portland (Oregon), San Francisco, Oakland, San Jose, Los Angeles, San

Diego, Galveston, Houston, New Orleans, Mobile, Tampa, St. Petersburg, Miami, Jacksonville, Savannah, Charleston, Wilmington, Norfolk, Washington, Baltimore, Philadelphia, New York, Providence, Boston, and Portland (Maine).

Other especially vulnerable nations that have large populations close to sea level include China, Brazil, the Netherlands, Singapore, Great Britain, Bangladesh, Indonesia, the Philippines, Denmark, India, and Japan. Many other nations, including Russia, Canada, Portugal, and Vietnam, have one or more major cities at risk. It would be folly to conclude that a "regional threat" does not concern us.

The cost of a survey that would provide 90% completeness of the population of 100-m bodies over a decade is higher than that envisioned for global killers because more sensitive detection systems must be used, and because the population to be discovered and monitored is 50 times larger than that of global killers. The cost of this option is likely to be about 10 times as large as the former, or about $50 million per year. The "premium" for this "insurance policy" is therefore about one-tenth the expected benefit. Adding in the cost of a 10^9 interception mission every 2000 years ($0.5 million per year) has no effect on this conclusion. A search program capable of finding 100-m NEOs is clearly a worthwhile investment.

LOCAL THREATS

We shall take as the threshold for severe local effects a body with a diameter of 10 m and an explosive yield of about 100 kt. The population of bodies in NEA orbits and in this size range is 2×10^8. Bodies in this yield range strike Earth about once a year, but most of them are fragile bodies that are destroyed by fragmentation while still at safe altitudes. The simulation runs presented in Chapter 7 show that lethal 10- to 100-kt impacts capable of devastating a town or small city occur at the rate of one every 100–200 years. The typical death toll in one such impact, which is usually a deeply penetrating airburst or surface impact of a rare, strong body, is about 1000 people. The annual death rate from these small bodies is therefore about 5–10 people per year, and the property loss is about $1 million per year.

The cost of discovering and tracking 100-m bodies already exceeds this figure. But the cost of monitoring the vastly larger population of 10-m bodies boggles the mind. The 10-m bodies are 2000 times as abundant as the 100-m bodies and 100 times smaller in cross-section area (five magnitudes fainter). The search and tracking effort required for the 10-m bodies therefore requires roughly 1000 times the sky coverage of the previous search, using telescopes of 10 times larger aperture. Even assuming economies of mass production, it seems most unlikely that such a search could be done for less than 100 to 1000 times the cost of the search for 100-m bodies. Identifying which of these bodies have physical properties that make them surface-impact threats (which ones are irons, stony-irons, or achondrites) involves further economic penalties, since these tiny bodies are so faint that they are almost never close enough for spectral characterization, and the prospect of sending spacecraft missions to characterize all of them is utterly out of the question. Thus the cost of targeting small (10-m) NEAs probably surpasses the expected benefits of finding them by roughly a factor of 10,000.

Hazard Abatement

Many techniques for deflecting or destroying threatening asteroids have been suggested. The most direct approach, and the most likely to be suggested first by newcomers, and by the authors of movie scripts and TV specials, is to "blow it up before it hits us." There are indeed some circumstances in which such as approach might be justified. For example, suppose that, contrary to probability, a comet was discovered on a collision trajectory with Earth only a few weeks before impact. Attempts to deflect the comet gently (i.e., without breakup) could not reach it quickly enough to deflect it far enough to ensure that it would miss Earth. In this case, "rubble-izing" the comet could, ideally, divert most of the fragments so that they would miss Earth and reduce the remainder to fragments small enough so that they would not be able to penetrate to Earth's surface. A devastating gigaton surface impact would be replaced by, say, fewer than a hundred 10-Mt blasts. If cometary nuclei are as fragile as some experts expect, these 10-Mt airbursts would occur at altitudes of 25–35 km, effectively sparing the surface from any damaging blast effects. The counterarguments that there may be, and indeed probably are, strong materials present inside comet nuclei, and that fragmentation to small enough pieces cannot be ensured, are both reasonable, but in these extreme circumstances an attempt at breaking up the body is the best option available. In any event, breaking up a very large body only hours before impact, when the fragments do not have time to drift clear of Earth impact trajectories, is certifiably suicidal. The area devastated by an explosion is proportional to $Y^{2/3}$. Therefore, if a 1000-Gt impactor is split into 1000 1-Gt bodies, the area devastated by their explosion would be $1000 \times 1^{2/3}$ compared to an area of $1 \times 1000^{2/3}$ for the intact large body, thus breaking it up into 1000 pieces *increases* the area affected by a factor of 10. The total energy content of the impact is conserved. Instead of a large area being subjected to overkill, an even larger area is killed.

Much more probable, and much easier to deal with, is the scenario in which an asteroid is predicted to make an approach to within a few Earth radii 200 years in the future. A series of gentle nudges, cumulating to a velocity change of only a few centimeters per second, could disturb the orbit enough to ensure a safe miss distance. Interestingly, there seem to be several viable ways to carry out such a deflection. Among the most attractive are standoff detonation of a nuclear device, deflection using a solar sail (Melosh and Nemchinov, 1993; Melosh *et al.*, 1994), and propulsion by a nuclear or solar steam rocket that uses surface materials of the asteroid or comet as its working fluid (Willoughby *et al.*, 1994). Electromagnetic propulsion systems, powered by solar electric or nuclear electric sources, are also possible. Such devices include rail guns and mass drivers.

9

Areas Requiring Further Study

The least satisfactory features of this model are the simplistic treatments of tsunamis, glass-related injuries, sulfur deposition, and atmospheric breakup. In view of the intrinsic complexity of the oceans and the immense variability in composition and morphology of the impactor population, there is very limited value to developing more sophisticated models of tsunami generation in isotropic plane-parallel oceans, or in studying the hydrodynamics of breakup of homogeneous viscous perfect spheres. Rather, it would seem desirable to expend some efforts on modeling the hydrodynamics of tsunami wave runup on several model shorelines that differ widely in their offshore topography, slope, and the tortuosity of their coastlines. Examples that come to mind include Los Angeles, Tokyo/Yokohama, Singapore, Bombay, Rio de Janiero, and Lisbon. Likewise, running several hydrodynamic models of the entry of bodies of highly nonspherical geometry or highly nonisotropic or scale-dependent physical properties could be instructive. Nonetheless, it is obvious that stochastic factors render predictions of any given breakup or tsunami event highly unreliable. Radar observations of near-Earth asteroids (Hudson and Ostro, 1994, 1995; Ostro *et al.*, 1995) provide convincing proof of their highly irregular shape and hint at highly nonuniform physical properties. Purely statistical models, which assume average properties and entry conditions, are capable of predicting only very long-term average results of these events.

The present methods of calculating the size of the 1-psi overpressure contour on the surface are both archaic and ridiculously cumbersome, being derived from engineering rules of thumb expressed in archaic units. Improvement in the method of calculation of the area affected by shock-wave accelerated window glass is long overdue.

Models of the atmospheric breakup and ablation of entering bodies have hitherto largely ignored the altitude dependence of deposition of sulfur, water, dust, and halogens in the upper atmosphere. Small fragile impactors, notably C asteroids and cometary debris, commonly deposit their energy and materials at altitudes in excess of 20 km, most of it in the immediate vicinity of the ozone layer. Such bodies must release virtually all of their sulfur as sulfur dioxide gas, the only form that is stable in an oxygen-rich atmosphere at high temperatures. Photochemical oxidation of atmospheric SO_2 leads directly to formation of sulfur trioxide gas, which rapidly adds a water vapor molecule to make sulfuric acid. Direct condensation of tiny, concentrated droplets of sulfuric acid, and reaction of sulfuric acid with the basic anhydride ammonia, is a normal feature of Earth's atmospheric chemistry, leading to the production and maintenance of the sulfate aerosols that constitute the stratospheric Junge layer. Greatly enhanced SO_2 abundances, such as those produced by volcanic eruptions or by bolide infall, can materially affect the density of the high-albedo Junge layer, and, in extreme cases (of order 10^{13} g of sulfur), brighten Earth sufficiently to cause detectable global cooling. Large fragile impactors, which penetrate deeper and generate buoyant plumes that lift sulfur oxides to high altitudes, are relatively simple to model (Kring *et al.*, 1996).

Several issues related to assessing the impact flux are obviously amenable to improved treatment. Technical progress in detectors and computerization of search programs has made it possible to carry out a greatly accelerated search for Earth-crossing bodies. Lessening of Cold War tensions has also resulted in the publication of a large body of data on fireballs derived from space-based infrared ballistic missile early-warning systems (Rawcliffe, 1979; Tagliaferri *et al.*, 1994). It is not clear that the full and optimal scientific use of these military assets has yet been achieved, and there has been no public discussion of the civil applications of future military space surveillance systems. A great increase in the discovery rate of near-Earth objects (NEOs) is clearly scientifically desirable, technically feasible, and relatively inexpensive (Rather *et al.*, 1992; Canavan *et al.*, 1994). Even at present discovery rates, however, the system for characterization of the near-Earth flux is very heavily burdened. High priority should be given to accelerating the physicochemical characterization of near-Earth bodies, whenever possible during their discovery opposition. Such remote-sensing data must be tested against ground truth established by flyby, rendezvous, and lander missions of small spacecraft sent to a compositionally and structurally diverse suite of NEOs.

It is difficult to address the threat posed by impacting bodies without a few words on the feasibility of interception and interdiction of these bodies. Some authors have argued that immediate construction and deployment of interceptors, bearing very large nuclear explosives, be initiated. Some have argued that true global threats are so infrequent that this is not a high-priority endeavor. Others, such as Carl Sagan (Sagan and Ostro, 1994; Harris *et al.*, 1994), have pointed out the possibilities for diversion of such vehicles for other, nefarious, purposes. These arguments greatly overestimate the ease of targeting an asteroidal impactor on Earth compared to simply ensuring a miss. I would argue that the discovery of a large (several hundred meter) threatening object is not enormously improbable; however, if we had a reasonably sensitive and thorough decade-long search program, we would typically detect the most threatening such body long before its actual impact. Then any threatening gigaton (10^4 year) object would, with a probability of 0.99, be detected at least 100 years before its threatened impact. This allows plenty of time to design, build, test, deploy, and use several generations of new-technology interceptors well before the event. Thus the most urgent present priority is to discover and characterize the near-Earth flux to ensure adequate warning times, not to build interceptor hardware.

Teller has argued (see, for example, Morrison and Teller, 1994) that, once a threatening object has been identified, the limiting factor in our response is uncertainty regarding the physical properties of the object and its response to external influences. Such factors as the density, porosity, crushing strength, heterogeneity, shape, and volatile content of the object are all important. For this reason, Teller recommends an early beginning to physical studies of Earth-crossing bodies.

We could, at a minimum, seek *in situ* data on the shape and internal physical properties of a single representative of each major compositional class of NEOs. We would then require rendezvous missions to at least one each of the following NEO classes: S, C, M, and V asteroids, short-period comets, and long-period comets. Many minor asteroid spectral classes, such as A and D, are also in evidence in the near-Earth population, and some of the traditional classes, such as M, may include members of two or more very different kinds of bodies with superficial spectral similarity (in the case of M asteroids, about half may be extremely tough metallic alloy, and the remainder may be water-bearing stony material of unknown affinity to established meteorite classes). Comet nuclei may be exceedingly idiosyncratic in their structure and physical strength, ranging from hard, brittle ice to porous fluff. Further reflection on the properties of meteorites suggests that the list of compositionally and structurally distinctive classes should include several dozen entries, the asteroidal sources of many of which are simply unknown. But consideration of the fruits of photometric and radar studies of a few fortuitously accessible NEAs suggests an even more unsettling conclusion: The shapes, structures, strengths, and rotation states of the members of even a single compositionally coherent class such as V asteroids are possibly, and even probably, endlessly diverse.

Such structural diversity in turn suggests that the responses of these objects to external forces, such as standoff nuclear explosions or landed propulsion systems, may also be extremely varied; indeed, the very same explosive yield delivered at the same distance from a rotating dumbbell-shaped "rubble pile" may have very different consequences depending on the orientation of the target at the time of detonation of the warhead. The idea that examination of a handful of varied targets will equip us with sufficient knowledge to plan interdiction missions is clearly naive. The implication of these considerations seems clear: Characterization, which is very desirable, leads us quickly into a regime in which large numbers of very inexpensive, mass-produced, highly autonomous spacecraft are required. This approach not only differs widely from the traditional NASA model, but even goes far beyond current concepts of $100-million missions. Rather than thinking of a single billion-dollar mission, or even 10 missions for the same total outlay, we should set our sights on having the ability to launch hundreds of missions with costs on the order of $1 to $5 million each. The full panoply of new technologies, from low-cost, fully reusable launch vehicles to novel launch techniques such as electromagnetic guns or slingatrons, highly competent onboards computers for navigation and data analysis, miniaturized low-power spacecraft instruments, and tiny sample-return vehicles would be required to achieve this goal at low total cost.

10

Conclusions

Earth, with a population density and distribution similar to that in the late 20th century, can expect an average of one or two fatal impact events per century. The statistical variability in the expected fatality rate is immense, since most deaths result from the one or two most severe events. At timescales of less than a few centuries, fatalities correlate poorly with impactor mass or energy. Rather, most fatal events are attributable to uncommon combinations of high impactor strength, low entry velocity, and low entry angle—statistical outriders of the impacting population. The bodies that threaten human civilization on a 100-year timescale are mostly 10- to 100-kt yield strong (iron, stony-iron, or achondrite) objects that liberate their energy very close to, or on, the surface. When only the average strength, velocity, and entry angle are used, these lethal events vanish. However, on the century timescale, the fatality rate is dominated by Tunguska-type airbursts of moderately strong chondritic material in the 1- to 100-Mt yield range, because these more energetic events are capable of affecting a much larger area near ground zero.

The leading cause of death in these simulations is fire, usually in the form of urban firestorms. Second in lethality are direct blast effects. In the size range typical of 100-year events, tsunamis are minor threats, but they become much more important on the 10,000-year timescale (i.e., over times comparable to the duration of human civilization). Glass-inflicted injury due to shattering and acceleration of window glass is common and devastating, but occurs mostly in areas subject to firestorms. Inadequacies of the computational model for glass-related injuries, however, suggest that this conclusion be treated with some caution.

It has been accepted for several years that most lethal impact events are caused by very small bodies drawn from a population that is extremely difficult to find and monitor. The present work finds that this principle is even more true than previously realized. A typical 30-kt yield iron has a diameter of only 9 m. Bodies of this size are 10^4 times as abundant as Tunguska-type 10-Mt impactors, and 10^6 times as common as kilometer-sized bodies. There are roughly a billion bodies of this size in near-Earth orbits, of which only about 3% are strong enough to be a serious local threat. It is certain that, given available and foreseeable technologies, it would not be cost effective to search for and track such an immense population of faint bodies merely to mitigate local threats.

Over timescales of 10,000 years, largest impactors in the 1000-Mt (1-Gt) range are expected. These bodies, which pose severe regional threats, are bright enough and limited enough in numbers so that a search and prediction program would be possible and economically sound. The crossover between unaffordable consequences and unaffordable remedies probably occurs at diameters of a little less than 100 m.

It would be especially useful to know precisely what the threshold size is for producing global-scale deleterious threats. It may be that the threshold size for affecting the global albedo with sulfate aerosols is near 300 m in diameter (1–2 Gt), but this issue is poorly understood. In any event, such events with global effects involve bodies that are large enough to make them relatively easy to discover and track. An active search program, capable as argued earlier of providing at least 100-year warnings of threatened impacts, would permit the leisurely development and testing of effective countermeasures. Our present circumstances would force us in an emergency to turn to very large nuclear devices for impactor interception and deflection (see, for example, Ahrens and Harris, 1992), but in the most likely scenarios we would have ample time to develop a non-nuclear interdiction capability (Melosh, 1994).

Clearly the most urgent item on our agenda must be the discovery of threatening large bodies. The recent acceleration of discovery activities has already provided a large increase in the discovery rate of NEOs. At a cost of only a few million dollars a year, the discovery rate could be further accelerated to a rate that would provide a census of NEOs that would be at least 90% complete in 20–30 years. This investment is easily justified in terms of the savings in damage costs and human lives. But the discovery rate is already so high that follow-up observations needed to ensure accurate orbital elements and to characterize the physical and chemical properties of potential impactors are lagging behind. Dedicated observatories, observers, and equipment are badly needed to provide us the physicochemical data needed for hazard assessment.

Of all the natural hazards facing Earth, impacts are the most dangerous. Unlike native hazards of Earth's surface, impacts know no size limit. Their effects can be devastating over the entire surface area of our planet. They are the only credible natural threat to human civilization. But impacts, especially those of large bodies, are both predictable and avoidable. The NEO population constitutes both an unprecedented hazard and an unparalleled opportunity. It is sometimes said that there is a fine line that separates a threat from an opportunity. The near-Earth asteroids present us with just this dilemma. They present us with an intelligence test of the highest order, with the highest possible stakes for the future of the human race.

APPENDIX

A

Program HAZARDS Version 5.5 Owner's Manual

Purpose

HAZARDS 5.5 is the newest version of a series of programs, which began (auspiciously) with HAZARDS 1.0 and has made continuous decimal progress since then. It models the comet and asteroid populations in the inner solar system and their most intimate interactions with the terrestrial planets. The physical and chemical properties and orbital statistics of the near-Earth asteroid, periodic comet, and long-period comet populations are all built in. The user has (in the full version) a choice of Mercury, Venus, Earth, and Mars as the target planet; other customized targets can be inserted without too much difficulty. Orbital perturbations and other long-term phenomena are ignored, and all interactions with planetary atmospheres and surfaces are modeled at least crudely.

MAIN PROGRAM
Organization

1. Identifies the version.
2. Calls INITIALIZE and FORMPRINT subroutines.
3. Loops through N time steps, calling nine subroutines as it cycles, and reporting on the biggest entry event in each time step.
4. Calls SUMMARY subroutine.
5. Offers choice of running another model.

Subroutine INITIALIZE

FUNCTIONS

1. Sets screen and printer parameters.
2. Clears bookkeeping registers.
3. Provides some handy constants.
4. Asks user to INPUT choice of target planet or specify new planet.
5. Reads in or calculates planetary gravitational and atmospheric parameters.
6. Asks user to INPUT size and number of time steps to be used.
7. Asks user to select printing, disk file, and screen display options.

Appendix A

8. Asks user whether to report all events or fatal events only.

Subroutine FORMPRINT

FUNCTIONS

1. Asks user to INPUT whether output should be sent to a printer.
2. Asks user to INPUT whether output should be sent to a disk file.
3. Displays/prints/saves output file header giving Ver. #, TIME, DATE, column heads.

Subroutine MASS

FUNCTIONS

1. Randomly calculates a mass for the impactor.

METHODS AND ASSUMPTIONS

1. Uses a statistical model of asteroid and comet mass data derived from astronomical observations of comets and belt asteroids ($>10^{16}$ g), lunar cratering statistics ($>10^{12}$ g), Spacewatch discovery statistics for NEAs (10^8–10^{12} g), and meteor fireball statistics (10^3–10^7 g).

Appendix A

Subroutine KIND

FUNCTIONS

1. Assigns asteroid or comet label based on size-frequency relation of near-Earth comet and asteroid populations.
2. Randomly assigns each asteroid a designation as a CI chondrite, CM chondrite, ordinary chondrite, achondrite, mesosiderite, pallasite, or iron, and each comet as either periodic or long period.
3. Generates data on the physical and chemical properties of the impactors.

METHODS AND ASSUMPTIONS

1. Uses statistical data on frequency of observed fall of meteorite classes, corrected for loss of C asteroidal material during entry (due to its low crushing strength) by use of ECAS spectral-class data on the NEA population, further corrected for observational bias against the discovery of low-albedo asteroids by techniques using visible light.
2. Uses laboratory physical and analytical data on the bulk composition, water content, and iridium content of meteorites and their density, crushing strength, enthalpy of vaporization, and luminous efficiency during entry.

Subroutine ENERGY

FUNCTIONS

1. Calculates velocities and kinetic energies for encounters with the chosen planet.
2. Randomly assigns an entry elevation angle.

Appendix A

METHODS AND ASSUMPTIONS

1. Uses a statistical model of the orbital encounter velocities of NEAs and comets.
2. Uses a fixed reference altitude (140 km for Earth, 220 km for Venus, 230 km for Mars) and a $\sin(2E)$ distribution of elevation angles at the reference altitude.

Subroutine BLOWOFF

FUNCTIONS

1. Gives advice to student users on how to avoid coursework.
2. Assesses whether explosive blowoff of atmosphere occurs and sets a flag.
3. Estimates the fraction of incident mass blown off.

METHODS AND ASSUMPTIONS

1. Uses Vickery/Melosh criteria to determine whether impactor is fast enough and energetic enough to cause atmospheric blowoff.

Subroutine FRAGMENT

FUNCTIONS

1. Analyzes the interaction of the entering body with the planetary atmosphere with an internal loop allowing up to 2000 time steps. The loop models the changes of the entry angle (for a spherical planet), mass, lumi-

nosity, and the vertical and horizontal components of velocity during entry deceleration, gravitational acceleration, and coordinate rotation, ablation, and breakup.
2. The following fates are recognized and flagged:
    a. burnup (ablation to less than 10^{-4} of the original mass)
    b. fragmentation (dynamic pressure overcomes internal strength)
    c. deceleration (to speeds too low for further ablation or fragmentation)
    d. impact (still intact at zero altitude; crater former)
    e. orbit (bolide exits atmosphere below escape speed)
    f. escape (bolide exits above escape speed)

METHODS AND ASSUMPTIONS

1. Simple Opik-type ablation model is appropriate for large bodies (viscous flow; air cap).
2. Crushing is treated as failure due to the stagnation-point pressure overcoming the crushing strength of the object; at present a Weibull-type size-dependent strength is implemented. Fragmentation is overridden if the object is too large or too low for the pieces to separate before impact.
3. Deceleration is tracked down only to 3 km s^{-1}, below which ablation is very slow.
4. All trajectory calculations are done for a round Earth, essentially by rotating the coordinate system used for a flat-Earth model after each time step. Skipout events are followed up to the initial injection altitude (140 km for Earth).

Appendix A

Subroutine CRATER

FUNCTIONS

1. Analyzes crater formation by impactors and high-mass fragmenters.
2. Reports yield (megatons), crater diameter, crater depth, area affected, and ejected mass.

METHODS AND ASSUMPTIONS

1. Uses a statistical model of cratering events for the target planet.

Subroutine HAZARD

FUNCTIONS (FOR EARTH IMPACTS ONLY)

1. Distinguishes between oceanic and continental impactors.
2. Assesses fatalities from impactor cratering.
3. Assesses tidal wave fatalities from oceanic impactors.
4. Assesses lethal and nonlethal blast injuries from airbursts.
5. Assesses fire hazard from fireball radiative heating.
6. Calculates mass and mixing ratio of NO_x produced by blast waves.

METHODS AND ASSUMPTIONS

1. Uses a statistical model of the population density distribution of Earth.

2. Uses nuclear-weapons-type models of blast wave effects and radiative heating.

Subroutine SORT

1. Bubble-sorts the impacts as they are generated to identify the three largest mass, the three highest iridium, and the three highest energy impactors ("3 Big 3").
2. Keeps running sums of these quantities.

Subroutine DATAPRINT

1. Manages screen display of (a subset of) the output data on each event.
2. Manages printing of (a larger subset of) the output data on each event.
3. Manages disk storage of the output data.

Subroutine SUMMARY

1. Manages screen display of the "3 Big 3" output from SORT.
2. Manages printing of the same output.
3. Manages disk storage of the same output.

APPENDIX

B

Program HAZARDS Version 5.5 Program Listing

The program is written so as to be capable of running under a variety of dialects of IBM Advanced Basic and GW Basic. Except for idiosyncrasies in the printing of certain graphics characters, the listing here is exact and, except for excision of code dealing with planets other than Earth (and therefore irrelevant to hazard estimation), complete.

Appendix B

```
10   VER=5.5 '                                                Program HAZARDS
20              GOSUB 230   '      subroutine INITIALIZE
30              GOSUB 990   '      subroutine FORMPRINT
40   FOR N=1 TO TSTEPS
50     EVENT=N: GOSUB 1450 '      subroutine MASS
60              GOSUB 1550 '      subroutine KIND
70              GOSUB 1910 '      subroutine ENERGY
80              GOSUB 2370 '      subroutine BLOWOFF
90              GOSUB 2610 '      subroutine FRAGMENT
100             GOSUB 3100 '      subroutine CRATER
110             GOSUB 3190 '      subroutine HAZARD
120             GOSUB 3670 '      subroutine SORT
130             GOSUB 4180 '      subroutine DATAPRINT
140  NEXT
150             GOSUB 4710 '      subroutine SUMMARY
160  COLOR 14: LOCATE 21,1: PRINT "É!!!!!!!!!!!!!!!!!!!!!!!!!!!!!!!!!»"
170  BEEP:              PRINT "º  Run Another Model? (y/n):      º"
180                     PRINT "È!!!!!!!!!!!!!!!!!!!!!!!!!!!!!!!!!¼"
190  LOCATE 22,30: INPUT "", A$
200  IF A$="Y" OR A$="y" THEN GOTO 20
210  CLOSE : END
220  '
230       '                  subroutine INITIALIZE
240  KEY OFF: CLS: COLOR 11
250  PRINT "Program HAZARDS Ver.";VER;" - - - Run at ";TIME$;" on ";DATE$;"."
260  PRINT "_____"
270  SUMIR=0 : SUME=0 : SUMM=0 : G=6.673E-08 : PI=3.14159 : SUNMASS=1.992E+33
280  FOR I=1 TO 3
290    MMASS(I)=0 : MEVENT(I)=0 : MTYPE$(I)="       " : MVIMP(I)=0
300    MEIMP(I)=0 : MMIR(I)=0   : MIRID(I)=0  : MFLAG$(I)="  "
310    IMASS(I)=0 : IEVENT(I)=0 : ITYPE$(I)="       " : IVIMP(I)=0
320    IEIMP(I)=0 : IMIR(I)=0   : IIRID(I)=0  : IFLAG$(I)="  "
330    EMASS(I)=0 : EEVENT(I)=0 : ETYPE$(I)="       " : EVIMP(I)=0
340    EEIMP(I)=0 : EMIR(I)=0   : EIRID(I)=0  : EFLAG$(I)="  "
350  NEXT
360  SUMBFAT=0: SUMFFAT=0: SUMTSFAT=0: SUMGFAT=0: SUMFATAL=0
370  IF A$="Y" OR A$="y" THEN GOTO 970 ELSE 380
380  PRINT "         É!!!!!!!!!!!!!!!!!!!!!!!!!!!!!!!!!!!!!!!»"
390  PRINT "         º      Select the Target Planet         º"
400  PRINT "         º      (Earth is the Default Option)    º"
410  PRINT "         ÇÄÄÄÄÄÄÄÄÄÄÄÄÄÄÄÄÄÄÄÄÄÄÄÄÄÄÄÄÄÄÄÄÄÄÄÄÄÄÄ¶"
420  PRINT "         º    1. Mercury   2. Venus   3. Earth   º"
430  PRINT "         º         4. Mars      5. other         º"
440  PRINT "         ÇÄÄÄÄÄÄÄÄÄÄÄÄÄÄÄÄÄÄÄÄÄÄÄÄÄÄÄÄÄÄÄÄÄÄÄÄÄÄÄ¶"
450  PRINT "         º         Selection (1-5):              º"
460  PRINT "         È!!!!!!!!!!!!!!!!!!!!!!!!!!!!!!!!!!!!!!!¼"
470  LOCATE 10,38: INPUT "", PLANET
```

Appendix B

```
480  PLANET=3  '  Other planet options suppressed
490  AU=1.49598E+13
540  PLANET$="Earth"  : MASS=5.974E+27 : RADIUS=6.378E+08  : Z0=120
550  RHOZERO=.001295  : H=580000!       : SUNDIST=1!*AU    : GOTO 840
830  SUNDIST= AU*SUND
840  PAREA=4*PI*RADIUS*RADIUS : DENSITY=3*MASS/(4*PI*RADIUS^3)
850  GRAV=G*MASS/(RADIUS*RADIUS) : VESCKM=SQR(2*G*MASS/RADIUS)/100000!
860  VORBKM=SQR(G*SUNMASS/SUNDIST)/100000!: COLOR 13: LOCATE 16,1
870  PRINT "É⌠⌠⌠⌠⌠⌠⌠⌠⌠⌠⌠⌠⌠⌠⌠⌠⌠⌠⌠⌠⌠⌠⌠⌠⌠⌠⌠⌠⌠⌠⌠⌠⌠⌠⌠⌠⌠⌠⌠»"
880  PRINT "º  Select Number and Size of Time Steps   º"
885  PRINT "º            Default is (0, 100)          º"
890  PRINT "ÇÄÄÄÄÄÄÄÄÄÄÄÄÄÄÄÄÄÄÄÄÄÄÄÄÄÄÄÄÄÄÄÄÄÄÄÄÄÄÄÄ¶"
900  PRINT "º  log (Time Step Size, Years):           º"
910  PRINT "º  Number of Time Steps:                  º"
920  PRINT "È⌠⌠⌠⌠⌠⌠⌠⌠⌠⌠⌠⌠⌠⌠⌠⌠⌠⌠⌠⌠⌠⌠⌠⌠⌠⌠⌠⌠⌠⌠⌠⌠⌠⌠⌠⌠⌠⌠⌠¼"
930  LOCATE 20,33: INPUT "", LOGTSTEP
940  LOCATE 21,26: INPUT "", TSTEPS
950  IF TSTEPS=0 THEN LOGTSTEP=0    ' set default size of time step = 1 yr
960  IF TSTEPS=0 THEN TSTEPS=100    ' set default number of time steps = 100
970  COLOR 14: RETURN
980  '
990  '            subroutine  FORMPRINT
1000 IF A$="Y" OR A$="y" THEN GOTO 1150 ELSE 1010
1010 LOCATE 17,40: PRINT "É⌠⌠⌠⌠⌠⌠⌠⌠⌠⌠⌠⌠⌠⌠⌠⌠⌠⌠⌠⌠⌠⌠⌠⌠⌠⌠⌠⌠⌠⌠⌠»"
1020 LOCATE 18,40: PRINT "º     Set Output Parameters     º"
1025 LOCATE 19,40: PRINT "º       (Default is no/no)      º"
1030 LOCATE 20,40: PRINT "ÇÄÄÄÄÄÄÄÄÄÄÄÄÄÄÄÄÄÄÄÄÄÄÄÄÄÄÄÄÄÄÄÄÄ¶"
1040 LOCATE 21,40: PRINT "º  Print Results? (y/n):        º"
1050 LOCATE 22,40: PRINT "º  Save Output to Disk? (y/n):  º"
1060 LOCATE 23,40: PRINT "È⌠⌠⌠⌠⌠⌠⌠⌠⌠⌠⌠⌠⌠⌠⌠⌠⌠⌠⌠⌠⌠⌠⌠⌠⌠⌠⌠⌠⌠⌠⌠¼"
1070 COLOR 10: LOCATE 21,65: INPUT "", P$
1090 LOCATE 22,71: INPUT "", D$
1095 IF P$="Y" OR P$="y" THEN GOTO 1097
1096 P$="n"
1097 IF D$="Y" OR D$="y" THEN GOTO 1110
1098 D$="n"
1110 LOCATE 16,10: PRINT "É⌠⌠⌠⌠⌠⌠⌠⌠⌠⌠⌠⌠⌠⌠⌠⌠⌠⌠⌠⌠⌠⌠⌠⌠⌠⌠⌠⌠⌠⌠⌠⌠⌠⌠⌠⌠⌠»"
1120 LOCATE 17,10: PRINT "º  Save fatal events only? (y/n):     º"
1125 LOCATE 18,10: PRINT "º            (Default is no)          º"
1130 LOCATE 19,10: PRINT "È⌠⌠⌠⌠⌠⌠⌠⌠⌠⌠⌠⌠⌠⌠⌠⌠⌠⌠⌠⌠⌠⌠⌠⌠⌠⌠⌠⌠⌠⌠⌠⌠⌠⌠⌠⌠⌠¼"
1140 LOCATE 17,43: COLOR 14: INPUT "", SHORT$
1145 IF SHORT$="Y" OR SHORT$="y" THEN GOTO 1150
1148 SHORT$="n"
1150 LOCATE 12,25: PRINT "É⌠⌠⌠⌠⌠⌠⌠⌠⌠⌠⌠⌠⌠⌠⌠⌠⌠⌠⌠⌠⌠⌠⌠⌠⌠⌠⌠⌠⌠⌠⌠⌠⌠⌠⌠⌠⌠»"
1160 LOCATE 13,25: PRINT "º                                    º"
1170 LOCATE 14,25: PRINT "º  Enter random number seed again:   º"
1180 LOCATE 15,25: PRINT "È⌠⌠⌠⌠⌠⌠⌠⌠⌠⌠⌠⌠⌠⌠⌠⌠⌠⌠⌠⌠⌠⌠⌠⌠⌠⌠⌠⌠⌠⌠⌠⌠⌠⌠⌠⌠⌠¼"
```

Appendix B

```
1190 LOCATE 13,27: COLOR 15: RANDOMIZE
1200 LOCATE 14,60: COLOR 14: INPUT "", RAN$
1210 COLOR 11: IF P$="n" OR P$="N" THEN GOTO 1300
1220 WIDTH "lpt1:",174:  LPRINT CHR$(27);"&l10"
1230 LPRINT "Program HAZARDS Ver.";VER;"- - -Run # ";RAN$; " at ";TIME$;
1240 LPRINT " on ";DATE$;" for ";PLANET$;": time step = 10^";LOGTSTEP; " years."
1250 LPRINT "Event Type       Mass    m(H2O) Velocity  Energy    Ir M/A  Fate Elev    Zmin  Vfinal  L(burn)  t(burn)  Max    ELfinal mass%   Yield  Xnox  Blowoff C rater  r(4psi)  LS"
1260 LPRINT "                  (g)         impact (km/s)  (erg)    (ug/cm2) B,F$ (deg)  (km)  (km/s)  (km)     (s)      mag    (deg)  left    Mton   (ppm)  FB     r (km)   (km)"
1270 LPRINT "    Offshore  runupHt  fldPen bFatal   fFatal  tsFatal  gFatal"
1280 LPRINT "      d (km)     (cm)    (km)"
1290 LPRINT
1300 IF D$="n" OR D$="N" THEN GOTO 1380
1310 IF A$="Y" OR A$="y" THEN GOTO 1340 ELSE 1320
1320 G$="out"+RAN$+".csv": OPEN G$ FOR OUTPUT AS #2
1330 PRINT#2,"Program HAZARDS Ver.";VER;"- -Run # ";RAN$;" at ";TIME$;" on ";DATE$;" for ";PLANET$;": time step = 10^";LOGTSTEP;" years."
1340 PRINT#2,"Event Type       Mass    m(H2O) Velocity  Energy    Ir M/A  B? I? Elev    Zmin Vfinal L(burn) t(burn)  Max    ELfinal mass%   Yield    Xnox   Blowoff Crater r(4psi) LS"
1350 PRINT#2,"Number            (g)         impact (km/s)  (erg)    (ug/cm2) B  F$ (deg)  (km)  (km/s)  (km)     (s)      mag    (deg)  left    Mton    (ppm)    FB    r(km)    (km)"
1360 PRINT#2," Offshore  runupHt  fldPen R(1psi)  Raff   FluExpl bFatal fFatal  ts Fatal  gFatal"
1370 PRINT#2," dist(km)  (cm)      (km)    (km)      (km)   erg/cm2 "
1380 CLS: PRINT "Event  Type     Mass Velocity Energy  Ir M/A    Zmin  Vfinal Fate  Eimp"
1390 PRINT "                 (g)   (km/s)  (erg)  (ug/cm2)  (km)  (km/s) B,F$  (Mton)"
1400 PRINT "    LS  r(4psi) Distance  runupHt  fldPen  bFatal    fFatal  tsFatal  gFatal"
1410 PRINT "        (km)    (km)      (cm)    (km)"
1420 PRINT
1430 RETURN
1440 '
1450 '               subroutine MASS
1460 L=LOGTSTEP+.2171*LOG(TSTEPS) : S=-.6: RND1=RND(1): IF L<4 THEN S=.15*L-1.2
1470 IF L<0 THEN S=-1.2
1480 IF L<-4 THEN S=-2.6667-.36667*L
1490 IF L<-5 THEN S=-.8333
1500 LT=LOGTSTEP
1510 Y=(10^(1.6667*(LT-2.2))+10^(.8333*(LT-1)))*RND1^(1/S)
1520 M=9E+12*Y/(30+1.67*.4343*LOG(Y))^2       '  correct for M-dependence of V
```

Appendix B

```
1530 RETURN
1540 '
1550 '          subroutine KIND
1560 C=RND(2): FLAG$= "  ": IF M>4.5E+16 THEN GOTO 1600
1570 IF C<.75 THEN TYPE$="asteroid"
1580 IF C>.75 THEN GOTO 1630
1590 GOTO 1640
1600 RATIO=10^(5.38-(.294*LOG(M)/2.303))
1610 IF M>10^18 THEN RATIO=1 : CFRACT=1-RATIO/(1+RATIO)
1620 IF CFRACT>RND(2) THEN GOTO 1630 ELSE GOTO 1640
1630 TYPE$="comet" : RETURN
1640 HVAP=8.08E+10: C=RND(2) : IF C<.082 THEN GOTO 1710
1650 R3=RND(3) : IF RND(3)<.77 THEN GOTO 1680
1660 ' measured ultimate strengths are: CI,1E7; CM,5E7; OC,1.7E9; ach/SI/I,3.7E9
1670 IR=.0000003:TYPE$="ast/CIch":FH2O=.2:RHO=2.25:CR=1E+07:TAU=.05: RETURN
1680 IF RND(3)<.72 THEN GOTO 1700
1690 IR=.0000004: TYPE$="ast/CMch": FH2O=.1:RHO=2.75:CR=5E+07:TAU=.05: RETURN
1700 IR=.0000005 : TYPE$="ast/Och ":FH2O=0!:RHO=3.61:CR=7E+08:TAU=.2: RETURN
1710 IF C<.031 THEN GOTO 1730
1720 IR=1E-08: TYPE$="ast/ach ": FH2O=0!: RHO=3.2: CR=1.7E+09:TAU=.02: RETURN
1730 IF C<.021 THEN GOTO 1770
1740 TAU=.3: IF RND(2)<.73 THEN GOTO 1760
1750 IR=.000002: TYPE$="ast/meso": FH2O=0!: RHO=5!: CR=3E+09: RETURN
1760 IR=3E-08: TYPE$="ast/pall": FH2O=0!: RHO=4.3+RND(5): CR=3E+09: RETURN
1770 IR=.0000035: TYPE$="ast/iron": FH2O=0!: RHO=7.9: CR=3.7E+09:TAU=.3:C=RND(2)
1780 HVAP=8.26E+10: IF C>.29 THEN IR=.0000015
1790 IF C>.46 THEN IR=.0000075
1800 IF C>.57 THEN IR=.000015
1810 IF C>.67 THEN IR=7.49E-07
1820 IF C>.74 THEN IR=3.5E-07
1830 IF C>.81 THEN IR=3.5E-08
1840 IF C>.86 THEN IR=.000035
1850 IF C>.9 THEN IR=1.5E-08
1860 IF C>.935 THEN IR=7.5E-08
1870 IF C>.97 THEN IR=1.5E-07
1880 IF C>.995 THEN IR=.000058
1890 RETURN
1900 '
1910 '          subroutine ENERGY
1920 IF LEFT$(TYPE$,3)="com" THEN GOTO 1940
1930 VIMP=100000!*(1.01*VESCKM+VORBKM*(1-COS(RND(2))^2)): GOTO 2320
1940 HVAP=3.2E+10: TAU=.02: ON PLANET GOTO 1950, 2040, 2130, 2220, 2310
1950 IF RND(2)>.5 GOTO 2000
1960 IR=.0000002 : TYPE$="com/per ": IF RND(5)>.5 THEN GOTO 1980 'Mercury impact
1970 VIMP=(29+23*RND(2)^2)*100000! : GOTO 1990
1980 VIMP=(29-23*RND(2)^2)*100000!
1990 FH2O=.5  : RHO=1.4  : CR=1E+08 : GOTO 2320
```

Appendix B

```
2000 IR=.0000001 : TYPE$="com/long" : IF RND(5)>.5 THEN GOTO 2020
2010 VIMP=(68+48*RND(2)^2)*100000! : GOTO 2030
2020 VIMP=(68-48*RND(2)^2)*100000!
2030 FH2O=.75 : RHO=1.1 : CR=1E+08 : GOTO 2320
2040 IF RND(2)>.5 GOTO 2090
2050 IR=.0000002 : TYPE$="com/per ": IF RND(5)>.5 THEN GOTO 2070   ' Venus impact
2060 VIMP=(26+14*RND(2)^2)*100000! : GOTO 2080
2070 VIMP=(26-14*RND(2)^2)*100000!
2080 FH2O=.5  : RHO=1.4 : CR=1E+08 : GOTO 2320
2090 IR=.0000001 : TYPE$="com/long" : IF RND(5)>.5 THEN GOTO 2110
2100 VIMP=(52+34*RND(2)^2)*100000! : GOTO 2120
2110 VIMP=(52-34*RND(2)^2)*100000!
2120 FH2O=.75 : RHO=1.1 : CR=1E+08 : GOTO 2320
2130 IF RND(2)>.5 GOTO 2180
2140 IR=.0000002 : TYPE$="com/per ": IF RND(5)>.5 THEN GOTO 2160   ' Earth impact
2150 VIMP=(24+11*RND(2)^2)*100000! : GOTO 2170
2160 VIMP=(24-11*RND(2)^2)*100000!
2170 FH2O=.5  : RHO=1.4 : CR=1E+08 : GOTO 2320
2180 IR=.0000001 : TYPE$="com/long" : IF RND(5)>.5 THEN GOTO 2200
2190 VIMP=(46+27*RND(2)^2)*100000! : GOTO 2210
2200 VIMP=(46-27*RND(2)^2)*100000!
2210 FH2O=.75 : RHO=1.1 : CR=1E+08 : GOTO 2320
2220 IF RND(2)>.5 GOTO 2270
2230 IR=.0000002 : TYPE$="com/per ": IF RND(5)>.5 THEN GOTO 2250   ' Mars impact
2240 VIMP=(16+10*RND(2)^2)*100000! : GOTO 2260
2250 VIMP=(16-10*RND(2)^2)*100000!
2260 FH2O=.5  : RHO=1.4 : CR=1E+08 : GOTO 2320
2270 IR=.0000001 : TYPE$="com/long" : IF RND(5)>.5 THEN GOTO 2290
2280 VIMP=(35+24*RND(2)^2)*100000! : GOTO 2300
2290 VIMP=(35-24*RND(2)^2)*100000!
2300 FH2O=.75 : RHO=1.1 : CR=1E+08 : GOTO 2320
2310 ' Planet X routine here
2320 MIR=M*IR : EIMP=M*VIMP^2/2 : IRID=MIR*100000!/PAREA : MH2OIMP=FH2O*M
2330 MTON=EIMP/(4.2E+22):Q=2*RND(4)-1: ELEV=28.6479*(1.570796-ATN(Q/SQR(1-Q*Q)))
2340 R=(3*M/(4*PI*RHO))^.333: IF EIMP>5E+34 THEN E$="e" ELSE E$=" "
2350 RETURN
2360 '
2370 '           subroutine BLOWOFF
2380 VI=VIMP/100000!: IF PLANET$="Mercury" THEN GOTO 2390 ELSE 2400
2390 FB=1: RETURN
2400 IF PLANET$="Venus" THEN GOTO 2410 ELSE 2420
2410 ' Venus FB routine here
2420 IF PLANET$="Mars" THEN GOTO 2430 ELSE 2440
2430 ' Mars FB routine here
2440 IF PLANET$="Earth" THEN GOTO 2450 ELSE 2520
2450 IF (LEFT$(TYPE$,3))="ast" THEN GOTO 2490
2460 IF LOG(M)/2.303>(27.5*LOG(VIMP)/2.303-160) THEN FB=.13333*VI-2.3
```

Appendix B

```
2470 IF LOG(M)/2.303<(27.5*LOG(VIMP)/2.303-160) THEN FB=-5.38+.2*LOG(EIMP)/2.303
2480 GOTO 2530
2490 IF LOG(M)/2.303>(20*LOG(VIMP)/2.303-110) THEN FB=.1*VI-1.75
2500 IF LOG(M)/2.303<(20*LOG(VIMP)/2.303-110) THEN FB=-7.63+.276*LOG(EIMP)/2.303
2510 GOTO 2530
2520 ' Planet X routine here
2530 IF FB>1 THEN FB=1
2540 IF FB<0! THEN FB=0!
2550 MIR=MIR*(1-FB)
2560 IRID=IRID*(1-FB) : MH2OKEPT=MH2OIMP*(1-FB) : IF FB=1 THEN FLAG$=" B "
2570 IF FB<1 THEN FLAG$=" b "
2580 IF FB=0 THEN FLAG$="   "
2590 RETURN
2600 '
2610 '          subroutine FRAGMENT
2620 F=0
2630 MO=M: MM=M: T=0: L=0: FC=1: EL=ELEV/57.2958 : IF RHOZERO=0 THEN RETURN
2640 V=VIMP:W=0: Z=100000!*ZO: ZMIN=Z: VZ=V*SIN(EL): VX=V*COS(EL): OLDM=0
2650 FOR ZSTEP=1 TO 2000
2660  R=(3*MM/(4*PI*RHO))^.333 : AREA=PI*R*R : CRS=CR*(4.4/MM)^.0833
2670  RHOATM=RHOZERO*EXP(-Z/H): DYNP=.5*RHOATM*V*V: DT=290000!*FC/V
2680  IF Z<300000! THEN DT=50000!*FC/V
2690  DV=-DYNP*AREA*DT/MM : DVZ=DV*SIN(EL)*DT/MM+GRAV*DT: DVX=DV*COS(EL)
2700  LAMBRHO=1E-08*(1.37+1.8*SQR(.00001*V)): IF V>2200000! THEN LAMBRHO=1.4E-07
2710  D=3*R*RHOATM/(4*LAMBRHO): GAMMA=.6/SQR(D)+.36/D: IF GAMMA>1 THEN GAMMA=1
2720  DM=-.413*GAMMA*AREA*RHOATM*DT*V^3/HVAP
2730  RAT=(DM/MM+DVX/VX+ABS(DVZ/VZ)) : IF RAT>-.3 THEN GOTO 2750
2740  FC=-.1*FC/RAT : GOTO 2670
2750  DZVEL=-(VZ+.5*DVZ)*DT: DZGRAV=-GRAV*DT*DT/2: DZGEOM=(VX*DT)^2/(RADIUS+Z)
2760  MM=MM+DM: DZ=DZVEL+DZGRAV+DZGEOM: VZ=-DZ/DT: V=SQR((V+DV)^2-GRAV*DZ)
2770  IF VZ>V THEN VZ=V
2780  VX=SQR(V*V-VZ*VZ): IF Z+DZ<Z THEN ZMIN=Z+DZ
2790  IF ZMIN<0 THEN ZMIN=0
2800  IF VX=0 THEN VX=.000001
2810  IF Z+DZ<0 THEN GOTO 2820 ELSE GOTO 2830
2820     DL=-DL*Z/DZ: DT=-DT*Z/DZ: DZ=-Z
2830  Z=Z+DZ: L=L+V*DT: T=T+DT: EL=ATN(VZ/VX): E=EL*57.2958: IF Z<0 THEN Z=0
2840  IF Z=0 THEN GOTO 2880
2850  F=-TAU*DM*V*V/(8*PI*DT*Z*Z): MAG=-26.7-1.0857*LOG(F/13000000!)
2860  IF MAG<OLDM THEN MMAX=MAG
2870  OLDM=MAG: W=.5*MM*V*V
2880   IF MM>.0001*MO THEN GOTO 2900
2890    MM=0! : W=0: F$=" b " : RETURN                    ' burned up
2900   IF V>300000! THEN GOTO 2940
2910    RRN=RND(4): IF RRN<.706 THEN F$="od "             ' decelerated
2920    IF RRN>.706 THEN F$="ld "
2930   RETURN
```

Appendix B

```
2940    IF DYNP<CRS THEN GOTO 3010
2950    F$=" f "                                                ' fragmented
2960    VDISP=1.9*V*SQR(RHOATM/RHO) : IF TBREAK=0 THEN TBREAK=T
2970    IF ZBREAK=0 THEN ZBREAK=Z
2980    R=R+VDISP*DT : AREA=PI*R*R : CRS=1.2*CR*(4.4/MM)^.0833
2990    IF W<.5*MO*VIMP*VIMP THEN GOTO 3000 ELSE 2670
3000    IF Z=0 THEN GOTO 3020 ELSE RETURN
3010    IF Z>0 THEN GOTO 3030
3020    F$=" i " : RETURN                                       ' impacted
3030    IF Z<100000!*ZO THEN GOTO 3070
3040    VESC=SQR(2*G*MASS/(RADIUS+ZO)): IF V<VESC THEN GOTO 3060
3050    F$=" e " : W=0! : RETURN                                ' escaped
3060    F$=" o " : W=0! : RETURN                                ' orbited
3070 NEXT ZSTEP
3080 RETURN
3090 '
3100 '           subroutine CRATER
3110 RCRATER=0: RAFF=0: AAFF=0: DCRATER=0: MEJECT=0: MTON=W/4.2E+22
3120 RFRAGSTR=H*SQR(RHOZERO/RHO)
3130 IF R>RFRAGSTR OR F$=" i " THEN GOTO 3140 ELSE RETURN
3140 RCRATER=4630!*(1000*MTON)^.3: RAFF=6*RCRATER: AAFF=PI*RAFF*RAFF
3150 DCRATER=2800*(1000*MTON)^.3: MEJECT=3.1416*RCRATER*RCRATER*2.6*DCRATER
3160 MDUST=.01*MEJECT
3170 RETURN
3180 '
3190 '           subroutine HAZARD
3200 IF PLANET=3 THEN GOTO 3210 ELSE RETURN
3210 FATAL=0: LS=-8: BFATAL=0: FFATAL=0: TSFATAL=0: GFATAL=0: LL=RND(4)
3220 LO=-9+8*LL^4 : SIGMAOCN=10^LO
3230 DISTKM=0: RUNUPHT=0: FLOODPEN=0: KK=RND(4): R4PSIKM=0: R1PSIKM=0
3240 IF F$=" e " OR F$=" b " OR F$=" o " THEN RETURN
3250 C=LOG(RND(3))/-.22: LS=10^(.1737*LOG(C)): SIGMALND=.2*10^LS ' pop. density
3260 YMT=W/4.185E+22: MNOX=1.3E-13*W
3264 IF YMT>100 THEN MNOX=30*10^(8+.8*LOG(YMT)/2.303)
3268 XNOX=1000000!*MNOX/(RHOZERO*H*PAREA) ' NOx
3270 ZKM=Z/100000!: YMT3=YMT^.333 : R4PSIKM=2.09*ZKM-.45*ZKM*ZKM/YMT3+5.08*YMT3
3280 R1PSIKM=10*YMT3*(1.7*ZKM^.5-1.36*ZKM^2+1.158)
3290 IF R1PSIKM<0 THEN R1PSIKM=0
3300 IF R4PSIKM<0 THEN R4PSIKM=0
3310 IF F$=" i " THEN GOTO 3330
3320 IF F$=" f " AND R>RFRAGSTR THEN GOTO 3330 ELSE GOTO 3440
3330 '               all Earth surface impacts
3340 RSLANT=1180000!*SQR(MTON)
3350 IF KK>.706 THEN GOTO 3430              ' separate ocean from land impacts
3360 F$="oi "                                                    ocean impacts
3370 AAFF=PI*R4PSIKM^2: BFATAL=.7*AAFF*SIGMAOCN        ' blast fatalities
3380 DISTKM=3657-5750000!*SQR(4.04496E-07-.0000004*RND(1)):DISTCM=100000!*DISTKM
```

Appendix B

```
3385 IF MTON>125 THEN GOTO 3388              '              assume depth = 5 km
3386 WVAMPCM=30400!*SQR(MTON)/DISTKM  :  GOTO 3390  '  deep water approximation
3388 WVAMPCM=101500!*SQR(SQR(MTON))/DISTKM       '  shallow water approximation
3390 RUNUPHT=30*WVAMPCM:IF RUNUPHT>150 THEN GOTO  3410 ELSE 3400
3400 FLOODPEN=0 : GOTO 3420
3410 FLOODPEN=10*(RUNUPHT-150)^(4/3)
3420 AFLOOD=DISTKM*FLOODPEN/100000!: TSFATAL=AFLOOD*SIGMALND: GOTO 3640
3430 F$="li ": AAFF=PI*R4PSIKM^2: BFATAL=AAFF*SIGMALND: GOTO 3580  ' land impacts
3440 IF F$=" f " AND R<RFRAGSTR AND Z>0 THEN GOTO 3450 ELSE RETURN
3450 ZKM=Z/100000!                                              ' air burst
3460 AAFF=PI*R4PSIKM^2                                   ' blast fatalities
3470 IF KK>.706 THEN BFATAL=AAFF*SIGMALND
3480 IF KK>.706 THEN F$="lf "
3490 IF KK<.706 THEN BFATAL=AAFF*SIGMAOCN
3500 IF KK<.706 THEN F$="of "
3510 IF KK<.706 THEN LS=LO
3520 TFIRE=.038*((1000*MTON)^.44)*(RHOATM/RHOZERO)^.36       ' fireball duration
3530 RSLANT=0: RAFF=0: AAFF=0: FFATAL=0
3540 FLUEXPL=4.4E+21*MTON/(PI*Z*Z): IF FLUEXPL<1E+09 THEN GOTO 3580
3550 RSLANT=ZKM*SQR(FLUEXPL/1E+09): RAFF=.7*SQR(RSLANT^2-ZKM^2)
3560 AAFF=PI*RAFF^2: FFATAL=.5*AAFF*SIGMALND            ' fire fatalities
3570 IF KK<.706 THEN FFATAL=.5*AAFF*SIGMAOCN
3580 GFATAL=.3*SIGMAOCN*R1PSIKM^2
3590 IF KK>.706 THEN GFATAL=.3*SIGMALND*R1PSIKM^2       ' glass fatalities
3600 IF FFATAL<.5 THEN FFATAL=0
3610 IF BFATAL<.5 THEN BFATAL=0
3620 IF TSFATAL<.5 THEN TSFATAL=0
3630 IF GFATAL<.5 THEN GFATAL=0
3640 FATAL=BFATAL+FFATAL+TSFATAL+GFATAL
3650 RETURN
3660 '
3670 '         subroutine SORT
3680 IF M<MMASS(3) THEN GOTO 3830
3690 IF M<MMASS(2) THEN GOTO 3700 ELSE GOTO 3720
3700 MEVENT(3)=EVENT: MMASS(3)=M: MVIMP(3)=VIMP: MEIMP(3)=EIMP: MMIR(3)=MIR
3710 MTYPE$(3)=TYPE$: MFLAG$(3)=FLAG$: MIRID(3)=IRID: GOTO 3830
3720 IF M<MMASS(1) THEN GOTO 3730 ELSE GOTO 3770
3730 MMASS(3)=MMASS(2) : MMASS(2)=M : MEVENT(3)=MEVENT(2): MIRID(3)=MIRID(2)
3740 MVIMP(3)=MVIMP(2): MEIMP(3)=MEIMP(2): MMIR(3)=MMIR(2): MFLAG$(3)=MFLAG$(2)
3750 MEVENT(2)=EVENT: MVIMP(2)=VIMP: MEIMP(2)=EIMP: MMIR(2)=MIR: MFLAG$(2)=FLAG$
3760 MTYPE$(3)=MTYPE$(2): MTYPE$(2)=TYPE$: MIRID(2)=IRID: GOTO 3830
3770 MMASS(3)=MMASS(2) : MMASS(2)=MMASS(1) : MMASS(1)=M : MEVENT(3)=MEVENT(2)
3780 MVIMP(3)=MVIMP(2): MEIMP(3)=MEIMP(2): MMIR(3)=MMIR(2): MFLAG$(3)=MFLAG$(2)
3790 MEVENT(2)=MEVENT(1): MVIMP(2)=MVIMP(1): MEIMP(2)=MEIMP(1): MMIR(2)=MMIR(1)
3800 MFLAG$(2)=MFLAG$(1): MFLAG$(1)=FLAG$: MIRID(3)=MIRID(2): MIRID(2)=MIRID(1)
3810 MVIMP(1)=VIMP: MEIMP(1)=EIMP: MMIR(1)=MIR: MIRID(1)=IRID
```

Appendix B

```
3820 MTYPE$(3)=MTYPE$(2): MTYPE$(2)=MTYPE$(1): MEVENT(1)=EVENT: MTYPE$(1)=TYPE$
3830 IF IRID<IIRID(3) THEN GOTO 3980
3840 IF IRID<IIRID(2) THEN GOTO 3850 ELSE GOTO 3870
3850 IEVENT(3)=EVENT: IMASS(3)=M: IVIMP(3)=VIMP: IEIMP(3)=EIMP: IMIR(3)=MIR
3860 IIRID(3)=IRID: ITYPE$(3)=TYPE$: IFLAG$(3)=FLAG$: GOTO 3980
3870 IF IRID<IIRID(1) THEN GOTO 3880 ELSE GOTO 3920
3880 IEVENT(3)=IEVENT(2):IMASS(3)=IMASS(2):IVIMP(3)=IVIMP(2):IEIMP(3)=IEIMP(2)
3890 IMIR(3)=IMIR(2): IIRID(3)=IIRID(2): IFLAG$(3)=IFLAG$(2): IEVENT(2)=EVENT
3900 IMASS(2)=M: IVIMP(2)=VIMP: IEIMP(2)=EIMP: IMIR(2)=MIR: IIRID(2)=IRID
3910 IFLAG$(2)=FLAG$: ITYPE$(3)=ITYPE$(2): ITYPE$(2)=TYPE$: GOTO 3980
3920 IEVENT(3)=IEVENT(2):IMASS(3)=IMASS(2):IVIMP(3)=IVIMP(2):IEIMP(3)=IEIMP(2)
3930 IMIR(3)=IMIR(2): IIRID(3)=IIRID(2): IFLAG$(3)=IFLAG$(2):IEVENT(3)=IEVENT(1)
3940 IMASS(2)=IMASS(1): IVIMP(2)=IVIMP(1): IEIMP(2)=IEIMP(1): IMIR(2)=IMIR(1)
3950 IIRID(2)=IIRID(1): IFLAG$(2)=IFLAG$(1): IEVENT(1)=EVENT: IMASS(1)=M
3960 IVIMP(1)=VIMP: IEIMP(1)=EIMP: ITYPE$(3)=ITYPE$(2): ITYPE$(2)=ITYPE$(1)
3970 ITYPE$(1)=TYPE$: IMIR(1)=MIR: IIRID(1)=IRID: IFLAG$(1)=FLAG$
3980 IF EIMP<EEIMP(3) THEN GOTO 4130
3990 IF EIMP<EEIMP(2) THEN GOTO 4000 ELSE GOTO 4020
4000 EEVENT(3)=EVENT: EMASS(3)=M: EVIMP(3)=VIMP: EEIMP(3)=EIMP : EMIR(3)=MIR
4010 ETYPE$(3)=TYPE$: EFLAG$(3)=FLAG$: EIRID(3)=IRID: GOTO 4130
4020 IF EIMP<EEIMP(1) THEN GOTO 4030 ELSE GOTO 4070
4030 EEVENT(3)=EEVENT(2):ETYPE$(3)=ETYPE$(2):EMASS(3)=EMASS(2):EVIMP(3)=EVIMP(2)
4040 EEIMP(3)=EEIMP(2): EMIR(3)=EMIR(2): EIRID(3)=EIRID(2): EFLAG$(3)=EFLAG$(2)
4050 EEVENT(2)=EVENT: ETYPE$(2)=TYPE$: EMASS(2)=M :EVIMP(2)=VIMP: EEIMP(2)=EIMP
4060 EMIR(2)=MIR: EIRID(2)=IRID: EFLAG$(2)=FLAG$: GOTO 4130
4070 EEVENT(3)=EEVENT(2): EMASS(3)=EMASS(2): EVIMP(3)=EVIMP(2):EEIMP(3)=EEIMP(2)
4080 ETYPE$(3)=ETYPE$(2):EMIR(3)=EMIR(2): EIRID(3)=EIRID(2): EFLAG$(3)=EFLAG$(2)
4090 EEVENT(2)=EEVENT(1): EMASS(2)=EMASS(1):EVIMP(2)=EVIMP(1): EEIMP(2)=EEIMP(1)
4100 EMIR(2)=EMIR(1): EIRID(2)=EIRID(1): EFLAG$(2)=EFLAG$(1):ETYPE$(2)=ETYPE$(1)
4110 EMASS(1)=M: EVIMP(1)=VIMP: EEIMP(1)=EIMP: EEVENT(1)=EVENT
4120 ETYPE$(1)=TYPE$: EMIR(1)=MIR: EIRID(1)=IRID: EFLAG$(1)=FLAG$
4130 SUMIR=SUMIR+MIR : SUME=SUME+EIMP : SUMM=SUMM+M
4140 SUMBFAT=SUMBFAT+BFATAL : SUMFFAT=SUMFFAT+FFATAL : SUMTSFAT=SUMTSFAT+TSFATAL
4150 SUMGFAT=SUMGFAT+GFATAL: SUMFATAL=SUMBFAT+SUMFFAT+SUMTSFAT
4160 RETURN
4170 '
4180 '          subroutine DATAPRINT
4190 IF P$="n" OR P$="N" THEN GOTO 4350
4195 IF SHORT$="N" OR SHORT$="n" THEN GOTO 4200
4196 IF FATAL<.5 THEN GOTO 4350
4200 LPRINT USING "#### "; EVENT;
4210 LPRINT USING "&"; TYPE$;
4220 LPRINT USING "##.##^^^^"; MO; MH2OIMP;
4230 LPRINT USING "###.## "; VIMP/100000!;
4240 LPRINT USING "##.##^^^^"; EIMP; IRID;
4250 LPRINT USING "&"; FLAG$; F$;
4260 LPRINT USING "##.##"; ELEV;
4270 LPRINT USING "####.## "; ZMIN/100000!; V/100000!; L/100000!; T; MMAX; E;
```

Appendix B

```
100 *MM/MO;
4280 LPRINT USING "##.##^^^^"; W/4.185E+22;
4290 LPRINT USING "####.#"; XNOX;
4300 LPRINT USING " #.####"; FB;
4310 LPRINT USING " ###.##"; RCRATER/100000!; R4PSIKM;
4320 LPRINT USING " ##.###"; LS
4330 LPRINT USING " ####.#"; DISTKM; RUNUPHT; FLOODPEN/100000!;
4340 LPRINT USING " ##.##^^^^"; BFATAL; FFATAL; TSFATAL; GFATAL
4350 IF D$="n" OR D$="N" THEN GOTO 4550
4360 IF SHORT$="N" OR SHORT$="n" THEN GOTO 4380
4370 IF FATAL<.5 THEN GOTO 4550
4380 PRINT#2,USING "####.,"; EVENT;
4390 PRINT#2,USING "&,"; TYPE$;
4400 PRINT#2,USING "##.##^^^^,"; MO; MH2OIMP;
4410 PRINT#2,USING "###.##,"; VIMP/100000!;
4420 PRINT#2,USING "##.##^^^^,"; EIMP;IRID;
4430 PRINT#2,USING "&,"; FLAG$;F$;
4440 PRINT#2,USING "##.##,"; ELEV;
4450 PRINT#2,USING "####.##,"; ZMIN/100000!; V/100000!; L/100000!; T; MMAX; E;
100 *MM/MO;
4460 PRINT#2,USING "##.##^^^^,"; W/4.185E+22;
4470 PRINT#2,USING "####.#,"; XNOX;
4480 PRINT#2,USING "#.###,"; FB;
4490 PRINT#2,USING "###.##,"; RCRATER/100000!; R4PSIKM;
4500 PRINT#2,USING "##.###,"; LS
4510 IF FATAL<.5 THEN GOTO 4550
4520 PRINT#2,USING "#####.##,"; DISTKM; RUNUPHT; FLOODPEN/100000!;
4530 PRINT#2,USING "#####.##,"; R1PSIKM; RAFF;
4540 PRINT#2,USING "####.##^^^^,"; FLUEXPL; BFATAL; FFATAL; TSFATAL; GFATAL
4550 IF SHORT$="N" OR SHORT$="n" THEN GOTO 4555
4552 IF FATAL<.5 THEN GOTO 4690
4555 PRINT USING "### "; EVENT;
4560 PRINT USING "&"; TYPE$;
4570 PRINT USING "##.##^^^^"; MO;
4580 PRINT USING "###.##"; VIMP/100000!;
4590 PRINT USING "##.##^^^^"; EIMP; IRID;
4600 PRINT USING "####.##"; ZMIN/100000!;
4610 PRINT USING "###.##"; V/100000!;
4620 PRINT USING "&"; FLAG$; F$;
4630 PRINT USING "##.##^^^^"; MTON
4640 IF FATAL>.5 THEN GOTO 4650 ELSE RETURN
4650 PRINT USING "##.###"; LS;
4660 PRINT USING " #####.#"; R4PSIKM; DISTKM; RUNUPHT; FLOODPEN/100000!;
4670 PRINT USING "########. "; BFATAL; FFATAL; TSFATAL; GFATAL
4680 PRINT F; MAG; TFIRE; FLUEXPL; RSLANT; AAFF
4690 RETURN
4700 '
```

```
4710 '              subroutine SUMMARY
4720 IF P$="n" OR P$="N" THEN GOTO 5050
4730 LPRINT "              Totals for model ";RAN$;":        ";
4740 LPRINT "    Sum(Mass) =";SUMM;"      Sum(Energy) =";SUME;"     Sum(Ir) =";SUMIR
4750 LPRINT : LPRINT "Big-3 Masses:"
4760 FOR J=1 TO 3
4770   LPRINT USING "####."; MEVENT(J);
4780   LPRINT USING "&"; MTYPE$(J);
4790   LPRINT USING "##.##^^^^"; MMASS(J);
4800   LPRINT USING "###.## "; MVIMP(J)/100000!;
4810   LPRINT USING "##.##^^^^"; MEIMP(J); MMIR(J); MIRID(J);
4820   LPRINT USING "&"; MFLAG$(J)
4830 NEXT
4840 LPRINT : LPRINT "Big-3 Iridium Events"
4850 FOR K=1 TO 3
4860   LPRINT USING "#### "; IEVENT(K);
4870   LPRINT USING "&"; ITYPE$(K);
4880   LPRINT USING "##.##^^^^"; IMASS(K);
4890   LPRINT USING "###.## "; IVIMP(K)/100000!;
4900   LPRINT USING "##.##^^^^"; IEIMP(K); IMIR(K); IIRID(K);
4910   LPRINT USING "&"; IFLAG$(K)
4920 NEXT
4930 LPRINT : LPRINT "Big-3 Energy Events"
4940 FOR L=1 TO 3
4950   LPRINT USING "#### "; EEVENT(L);
4960   LPRINT USING "&"; ETYPE$(L);
4970   LPRINT USING "##.##^^^^"; EMASS(L);
4980   LPRINT USING "###.## "; EVIMP(L)/100000!;
4990   LPRINT USING "##.##^^^^"; EEIMP(L); EMIR(L); EIRID(L);
5000   LPRINT USING "&"; EFLAG$(L)
5010 NEXT
5020 LPRINT
5025 LPRINT "Fatalities for model ";RAN$;"(B/F/Ts/G): ";
5030 LPRINT "sumBfat = ";SUMBFAT;"sumFfat = ";SUMFFAT;"sumTSfat = ";SUMTSFAT;"sumGfat = ";SUMGFAT
5040 LPRINT CHR$(12)
5050 PRINT "              Sums for run ";RAN$;":"
5060 PRINT"Sum(Mass) =";SUMM;"   Sum(Energy) =";SUME;"   Sum(Ir) ="; SUMIR
5070 PRINT : PRINT "Big-3 Masses:"
5080 FOR J=1 TO 3
5090   PRINT USING "#### "; MEVENT(J);
5100   PRINT USING "&"; MTYPE$(J);
5110   PRINT USING "##.##^^^^"; MMASS(J);
5120   PRINT USING "###.## "; MVIMP(J)/100000!;
5130   PRINT USING "##.##^^^^"; MEIMP(J); MMIR(J); MIRID(J);
5140   PRINT USING "&"; MFLAG$(J)
```

Appendix B

```
5150 NEXT
5160 PRINT : PRINT "Big-3 Iridium Events:"
5170 FOR K=1 TO 3
5180   PRINT USING "#### ";  IEVENT(K);
5190   PRINT USING "&"; ITYPE$(K);
5200   PRINT USING "##.##^^^^"; IMASS(K);
5210   PRINT USING "###.## "; IVIMP(K)/100000!;
5220   PRINT USING "##.##^^^^"; IEIMP(K); IMIR(K); IIRID(K);
5230   PRINT USING "&"; IFLAG$(K)
5240 NEXT
5250 PRINT : PRINT "Big-3 Energy Events:"
5260 FOR L=1 TO 3
5270   PRINT USING "#### "; EEVENT(L);
5280   PRINT USING "&"; ETYPE$(L);
5290   PRINT USING "##.##^^^^"; EMASS(L);
5300   PRINT USING "###.## "; EVIMP(L)/100000!;
5310   PRINT USING "##.##^^^^"; EEIMP(L); EMIR(L); EIRID(L);
5320   PRINT USING "&"; EFLAG$(L)
5330 IF E$="e" THEN GOTO 5340 ELSE GOTO 5370
5340 PRINT "                      Congratulations!"
5350 PRINT"You have just boiled away the oceans, destroying all life on Earth!"
5360 PRINT "                      END OF GAME"
5370 NEXT
5375 PRINT "Fatalities for model ";RAN$;" (B/F/Ts/G): ";
5380 PRINT"sumBfat = ";SUMBFAT;"sumFfat = ";SUMFFAT;"sumTSfat = ";SUMTSFAT;"sumGfat = ";SUMGFAT
5390 PRINT: PRINT: PRINT: IF D$="n" OR D$="N" THEN RETURN
5400 PRINT#2,"Big-3 Masses:,"
5410 FOR J=1 TO 3
5420   PRINT#2,USING "####.,"; MEVENT(J);
5430   PRINT#2,USING "&,"; MTYPE$(J);
5440   PRINT#2,USING "##.##^^^,"; MMASS(J);
5450   PRINT#2,USING "###.##,"; MVIMP(J)/100000!;
5460   PRINT#2,USING "##.##^^^,"; MEIMP(J); MMIR(J); MIRID(J);
5470   PRINT#2,USING "&,"; MFLAG$(J);
5480 NEXT
5490 PRINT#2,"Big-3 Iridium Events:,"
5500 FOR K=1 TO 3
5510   PRINT#2,USING "####.,"; IEVENT(K);
5520   PRINT#2,USING "&,"; ITYPE$(K);
5530   PRINT#2,USING "##.##^^^,"; IMASS(K);
5540   PRINT#2,USING "###.##,"; IVIMP(K)/100000!;
5550   PRINT#2,USING "##.##^^^,"; IEIMP(K); IMIR(K); IIRID(K);
5560   PRINT#2,USING "&,"; IFLAG$(K);
5570 NEXT
5580   PRINT#2,"Big-3 Energy Events:,"
5590 FOR L=1 TO 3
```

```
5600 PRINT#2,USING "####.,"; EEVENT(L);
5610 PRINT#2,USING "&,"; ETYPE$(L);
5620 PRINT#2,USING "##.##^^^^,"; EMASS(L);
5630 PRINT#2,USING "###.##,"; EVIMP(L)/100000!;
5640 PRINT#2,USING "##.##^^^^,"; EEIMP(L); EMIR(L); EIRID(L);
5650 PRINT#2,USING "&,"; EFLAG$(L);
5660 NEXT
5665 PRINT#2,"Fatalities for model ";
5666 PRINT#2, RAN$;
5667 PRINT#2," (B,F,Ts,G): ";
5670 PRINT#2,USING "##.##^^^^,";SUMBFAT;SUMFFAT;SUMTSFAT;SUMGFAT
5680 RETURN
```

APPENDIX

C

Program HAZARDS Version 5.5 Sample Output

Program output is in three forms: display to screen (obligatory) of important highlights of the results, save to disk in comma-separated-variable format (*.csv), a form readily imported by almost all database and word processing programs (optional), and direct print (also optional). My practice is to save all runs to disk and prepare printouts from the *.csv files by simply setting appropriate margins and specifying landscape mode printing.

For archival purposes, every run begins with a statement of the program and version number, the time and date, the planet (always Earth in this simplified program), and the size of the time step specified at the start of the run.

The headings on the columns of output are as follows:

Appendix C

Event Number: The number of the time step in which the event occurred. Only the largest energetic event is reported in each time interval. Experience has shown that this simplification results in a negligible underreporting of fatalities.

Type: The compositional class to which this object belongs. All are specified as asteroidal (ast) or cometary (com), followed by the detailed type. Only two types of comets are recognized because our limited body of compositional data on comets does not justify more detailed description. These two are long-period comets (com/long) and short-period or periodic comets (com/per). The asteroid types are based on studies of meteorites, and surely are an incomplete characterization of the full range of asteroidal bodies. The asteroidal types are Ivuna-type carbonaceous chondrites (CIch), Murchison-type carbonaceous chondrites (CMch), ordinary chondrites (Och), achondrites (ach), mesosiderites (meso), pallasites (pall), and irons (iron).

Mass (g): The entry mass of the body as it passes the 140-km altitude mark, in grams.

m(H2O) impact: The total mass of water delivered by the impactor. This is important for Mars and Venus simulations, but has only a secondary importance on Earth: In the event of large injections of sulfur, the background water content of the stratosphere may be insufficient to convert all of the sulfur oxides to sulfuric acid.

Velocity (km/s): The entry velocity of the body at the 140-km level. This is compounded from two contributions, the kinetic energy of the body in free orbit about the Sun and the gravitational potential energy contributed by the fall of the body from infinity to 140 km in Earth's gravity field.

Energy (erg): the kinetic energy of the impactor as it passes through the 140-km altitude.

Appendix C

Ir M/A (ug/cm²): The globally averaged area loading of iridium injected by the impactor in micrograms per square centimeter. This parameter is of interest principally for very large bodies.

B? and I?: These parameters code the fate of the entering body. The first column flags bodies that burn up during entry (b). These bodies obviously do not appear in a list of fatal impact events. The first letter in the second column indicates whether the even occurred over land (l) or ocean (o). The final flag denotes fragmentation (f), deceleration to below 3 km s^{-1} (d), impact (i), skipout to escape (e), and skipout to orbit (o).

Elev (deg): Gives the entry angle relative to the horizontal. Vertical incidence is therefore 90°. All numbers are, of course, negative; the minus sign is omitted to save space.

Zmin (km): For bodies that burn up or fragment catastrophically, Zmin is the altitude at which they last broke up. For some very large bodies that break up into a compact swarm, the swarm may be dense enough to permit surface impact. In these rare cases, Zmin is not zero but the fate is recorded as an impact (i).

Vfinal (km/s): This is the terminal velocity of the body when calculation of its trajectory ceased (at burnup, fragmentation, deceleration to less than 3 km s^{-1}, or skipout through 140-km altitude).

L(burn) (km): This is the length in kilometers of the chord below 140 km over which the body flew as a luminous, intact fireball.

t(burn) (s): This is the duration of burn, as defined earlier, in seconds.

Max mag: This is the peak brightness of the fireball in stellar magnitudes as seen from the ground directly beneath the fireball. For comparison, the magnitude of the Sun as seen from Earth is −26.4. Magnitudes are expressed on a logarithmic scale in which every factor of 100 in luminosity corresponds to exactly 5 magnitudes. The brighter the source, the more negative the magnitude. Note that some of the brightest fireballs can provide more than 100 times the intensity of full sunlight.

ELfinal (deg): The elevation angle of the trajectory of the body at the end of the integration of its path (see Vfinal earlier), taking into account deceleration, gravity, and rotation of the coordinate system. For bodies that skip out, the number will be negative.

mass% left: The percentage of the original entering mass that survives at the end of trajectory integration. The number may range from 100.00% for very massive bodies that suffer little ablation to 0.00% for bodies that burn up completely.

Yield Mton: The explosive yield (kinetic energy content) of the body at the time of fragmentation or impact, in megatons of TNT (4.2×10^{22} erg per megaton). The megaton is not a metric unit (tons here are English tons), but also not precisely an English unit, because the explosive yield of 1 ton of TNT has been rounded to 10^{15} calories.

Xnox (ppm): The globally averaged concentration of total nitrogen oxides (initially NO, but later NO_2 and its dimer N_2O_4). This number is negligible for 10,000-year events. However, NO may be important locally after an impact.

Blowoff FB: The fraction of Earth's atmosphere blown off by an impact event. This is generally negligible for million-year events and smaller.

Crater r(km): The radius of the transient crater produced by a surface impact in water or on land.

r(4psi) (km): The radius of the region on the ground that is shocked to blast overpressures in excess of 4 psi (0.25 atm). This overpressure is sufficient to destroy all except specially hardened construction.

Appendix C

LS: Log sigma, the log to base 10 of the population density in the target area, in people per square kilometer. In a list of major fatal events, one rarely sees small values of LS. In a complete listing of events, one encounters 10- to 100-Mt events that cause no fatalities because they occur in very sparsely populated areas.

Offshore dist(km): For ocean impacts, the distance of the impact point from the nearest inhabited coastline. This statistic is generated from population distribution and geographical data.

runupHt (cm): This is the height to which a tsunami wave runs up on the shore after propagating from the offshore site of impact. Typically the runup is about a factor of 20 or 30 over the height of the open-ocean wave when it first encounters land.

fldPen (km): This is the inland distance of penetration of flooding from tsunami waves that run up on a coastline.

R(1psi) (km): The radius of the region on the surface that experiences blast overpressures in excess of 1 psi, a number sufficient to shatter windows and accelerate glass to high velocities. The calculation of this parameter is not very satisfactory in current models.

Raff (km): The radius of the area affected by blast or fire effects.

FluExpl erg/cm2: The radiant energy fluence (flux integrated over time) for a point directly below an aerial explosion. The number is estimated from nuclear weapons codes.

bFatal: The number of fatalities due to blast effects.

fFatal: The number of fatalities due to firestorm ignition.

tsFatal: The number of fatalities due to tsunami inundation.

gFatal: The number of fatalities due to accelerated glass.

```
Program HAZARDS Ver. 5.5  - Run # 3000 at 18:38:11 on 05-26-1998 for Earth; time step = 10^0 years.
Event Type    Mass    Velocity  Energy    Ir M/A  B7  I?  Elev     Zmin  Vfinal L(burn) t(burn)    Max    Elfinal  mass%  Yield  Xnox   Blowoff Crater  r(4psi)  LS
Number        (g)     Impact    (erg)     (ug/cm2) B   F$  (deg)   (km)  (km/s)  (km)    (s)        mag    (deg)    left   Mton   (ppm)  FB      r(km)   (km)
Offshore runupHt fidPen R(psi)  Raff       FluExpl bFatal fFatal  tsFatal gFatal
dist(km) (cm)  (km)    (km)     erg/cm2
   4., ast/Och ,   3.68E+10,  0.00E+00, 25.00, 0.00, 1.15E+23, 3.60E-10,   ,lf , 69.64, 16.73,  24.96, 110.19,   4.41, -25.36, 69.87, 99.97, 2.74E+00, 0.0, 0.000,  0.00,  0.00, 1.125,
   0.00,   0.00,   0.00,   0.00,   7.09, 136.59E+07,   0.00E+00, 210.54E+00,   0.00E+00,   0.00E+00,
  13., ast/iron,   1.18E+09,  0.00E+00, 15.54,  3.84E+21, 2.18E-11,   ,lf , 44.61,  2.27, 14.65, 168.03, 10.84, -26.65, 47.55, 99.90, 8.16E-02, 0.0, 0.000, 0.00, 1.60, 3.085,
   0.00,   0.00,   0.00,   0.00,   1.75, 220.44E+07, 195.85E+01, 116.56E+01,   0.00E+00,   0.00E+00,
  97., ast/Och ,   6.16E+10,  0.00E+00, 14.79,  6.74E+22, 6.02E-10,   ,lf , 43.56, 12.64, 14.73, 156.59, 10.58, -23.55, 43.33, 99.99, 1.60E+00, 0.0, 0.000, 0.00, 0.00, 1.590,
   0.00,   0.00,   0.00,   0.00,   5.56, 139.46E+07,   0.00E+00, 377.62E+00,   0.00E+00,   0.00E+00,
 123., ast/Och ,   4.93E+11,  0.00E+00, 19.74,  9.60E+23, 4.82E-09,   ,lf , 76.90, 15.45, 19.74, 107.30,  5.43, -25.55, 77.38, 100.00, 2.29E+01, 0.0, 0.000, 0.00, 8.85, 2.636,
   0.00,   0.00,   0.00,   0.00,  33.09, 134.03E+08, 213.09E+02, 197.30E+03,   0.00E+00,   0.00E+00,
 141., ast/CMch,   3.99E+11,  0.00E+00, 21.31,  9.06E+23, 3.12E-09,   ,lf , 44.48, 25.09, 21.31, 136.30,  6.39, -23.72, 43.85, 99.99, 2.16E+01, 0.0, 0.000, 0.00, 0.00, 1.739,
   0.00,   0.00,   0.00,   0.00,  34.23, 479.86E+07,   0.00E+00, 201.86E+02,   0.00E+00,   0.00E+00,
 205., ast/Och ,   2.09E+11,  0.00E+00, 17.82,  3.33E+23, 2.05E-09,   ,lf , 26.48, 16.74, 17.79, 237.79, 13.33, -24.36, 25.10, 99.99, 7.91E+00, 0.0, 0.000, 0.00, 0.00, 2.035,
   0.00,   0.00,   0.00,   0.00,  20.09, 393.99E+07,   0.00E+00, 137.44E+02,   0.00E+00,   0.00E+00,
 210., ast/Och ,   5.58E+10,  0.00E+00, 11.76,  3.86E+22, 5.46E-10,   ,lf , 38.74,  9.82, 11.66, 176.87, 15.02, -23.01, 38.82, 99.99, 9.06E-01, 0.0, 0.000, 0.00, 0.00, 1.720,
   0.00,   0.00,   0.00,   0.00,   3.84, 131.11E+07,   0.00E+00, 242.72E+00,   0.00E+00,   0.00E+00,
 230., ast/iron,   1.29E+10,  0.00E+00, 20.74,  2.77E+22, 8.83E-11,   ,lf , 65.50,  3.90, 20.33, 127.54,  6.15, -27.85, 68.08, 99.94, 6.37E-01, 0.0, 0.000, 0.00, 4.57, 1.953,
   0.00,   0.00,   0.00,   0.00,   6.01, 583.88E+07, 117.52E+01, 101.73E+01,   0.00E+00,   0.00E+00,
 243., ast/Och ,   9.04E+10,  0.00E+00, 26.24,  3.11E+23, 8.84E-10,   ,lf , 79.64, 17.28, 26.21, 104.40,  3.98, -25.98, 80.27, 99.98, 7.42E+00, 0.0, 0.000, 0.00, 0.00, 0.248,
   0.00,   0.00,   0.00,   0.00,  19.00, 346.80E+07,   0.00E+00, 200.58E+00,   0.00E+00,   0.00E+00,
 282., ast/iron,   5.14E+09,  0.00E+00, 20.63,  1.10E+22, 1.51E-10,   ,lf , 53.19,  4.18, 20.01, 144.91,  7.03, -27.12, 55.17, 99.89, 2.46E-01, 0.0, 0.000, 0.00, 0.00, 1.504,
   0.00,   0.00,   0.00,   0.00,   2.87, 196.11E+07,   0.00E+00, 826.14E+01,   0.00E+00,   0.00E+00,
 443., ast/Och ,   2.63E+09,  0.00E+00, 14.78,  2.87E+21, 1.80E-12,   ,lf , 73.58,  2.39, 14.18, 122.19,  8.28, -26.34, 90.00, 99.93, 6.31E-02, 0.0, 0.000, 0.58, 2.679,
   0.00,   0.00,   0.00,   0.00,   1.23, 154.39E+07, 992.81E-01, 227.84E+00,   0.00E+00,   0.00E+00,
 466., com/long,   9.36E+10,  7.02E+10, 61.77,  1.79E+24, 1.83E-10,   ,lf , 27.84, 41.86, 61.77, 171.10,  2.77, -25.73, 26.52, 99.85, 4.26E+01, 0.0, 0.000, 0.00, 0.00, 3.109,
   0.00,   0.00,   0.00,   0.00,  45.32, 339.22E+07,   0.00E+00, 829.79E+03,   0.00E+00,   0.00E+00,
 499., com/long,   2.16E+11,  1.62E+11, 33.85,  1.24E+24, 4.22E-10,   ,lf , 52.67, 32.70, 33.85, 110.20,  3.26, -24.21, 52.15, 99.97, 2.95E+01, 0.0, 0.000, 0.00, 0.00, 0.972,
   0.00,   0.00,   0.00,   0.00,  38.67, 385.47E+07,   0.00E+00, 440.81E+01,   0.00E+00,   0.00E+00,
 532., ast/iron,   2.91E+09,  0.00E+00, 12.99,  2.45E+21, 8.53E-11,   ,oi, 10.10,  0.00,  7.30, 866.39, 69.09, -30.64, 10.12, 99.62, 1.87E-02, 0.0, 0.000, 0.11, 1.35, 0.957,
  71.48, 1740.03,   0.00,   3.08, 66771.40, 239.48E+05, 853.23E-11,   0.00E+00, 240.06E+00,   0.00E+00,
 576., ast/iron,   2.12E+09,  0.00E+00, 13.49,  1.93E+21, 2.40E-10,   ,lf , 34.84,  1.24, 12.17, 209.02, 15.56, -26.92, 38.41, 99.88, 3.74E-02, 0.0, 0.000, 0.00, 2.23, 0.927,
   0.00,   0.00,   0.00,   3.25,   1.35, 341.87E+07, 264.63E-01, 481.13E-02,   0.00E+00,   0.00E+00,
 604., ast/ach ,   6.05E+09,  0.00E+00, 11.58,  4.05E+21, 1.18E-12,   ,lf , 69.21,  2.98, 11.00, 124.55, 10.77, -22.24, 82.37, 99.96, 8.74E-02, 0.0, 0.000, 0.00, 0.00, 0.375,
   0.00,   0.00,   0.00,   0.00,   1.27, 137.12E+07,   0.00E+00, 536.18E-02,   0.00E+00,   0.00E+00,
 661., com/long,   8.94E+10,  6.70E+10, 46.53,  9.68E+23, 1.75E-10,   ,of , 37.07, 37.07, 46.53, 139.20,  2.99, -24.83, 36.11, 99.90, 2.31E+01, 0.0, 0.000, 0.00, 0.00, -1.698,
   0.00,   0.00,   0.00,   0.00,  30.10, 234.55E+07,   0.00E+00, 285.10E-01,   0.00E+00,   0.00E+00,
 734., ast/Och ,   1.59E+11,  0.00E+00, 11.81,  1.11E+23, 1.55E-09,   ,lf , 36.45, 10.43, 11.75, 185.59, 15.69, -23.44, 36.23, 99.99, 2.62E+00, 0.0, 0.000, 0.00, 0.00, 1.880,
   0.00,   0.00,   0.00,   0.00,  11.21, 335.76E+07,   0.00E+00, 299.37E+01,   0.00E+00,   0.00E+00,
 771., ast/Cich,   2.33E+11,  4.65E+10, 25.70,  7.68E+23, 1.37E-09,   ,lf , 72.24, 26.12, 25.69, 98.60,  3.83, -24.60, 72.34, 99.99, 1.83E+01, 0.0, 0.000, 0.00, 0.00, 2.764,
   0.00,   0.00,   0.00,   0.00,  30.33, 375.11E+07,   0.00E+00, 167.92E+03,   0.00E+00,   0.00E+00,
 787., ast/Och ,   3.54E+10,  0.00E+00, 22.31,  8.82E+22, 3.47E-10,   ,lf , 63.46, 16.34, 22.27, 115.99,  5.20, -24.82, 63.52, 99.97, 2.10E+00, 0.0, 0.000, 0.00, 0.00, 2.243,
   0.00,   0.00,   0.00,   0.00,  22.33, 481.87E+07, 264.63E-01, 694.12E+00,   0.00E+00,   0.00E+00,
 819., ast/Och ,   1.33E+11,  0.00E+00, 30.49,  6.17E+23, 1.30E-09,   ,lf , 83.25, 19.18, 30.47, 101.50,  3.33, -26.59, 83.96, 99.98, 1.47E+01, 0.0, 0.000, 0.00, 0.00, 1.574,
   0.00,   0.00,   0.00,   0.00,  28.74, 558.30E+07,   0.00E+00, 973.98E+01,   0.00E+00,   0.00E+00,
Big-3 Masses:,                505., com/long , 6.95E+11,  1.04E+25, 54.81, 1.04E+25,  6.95E+04, 1.56E-09,
                              698., ast/Cich , 6.24E+11,  2.30E+24, 27.16, 2.30E+24,  1.87E+05, 3.66E-09,
Big-3 Iridium Events:,        123., ast/Och  , 4.93E+11,  9.60E+23, 19.74, 9.60E+23,  2.46E+05, 4.82E-09,
                              698., ast/Cich , 6.24E+11,  2.30E+24, 27.16, 2.30E+24,  1.87E+05, 3.66E-09,
                              141., ast/CMch , 3.99E+11,  9.06E+23, 21.31, 9.06E+23,  1.60E+05, 3.12E-09,
Big-3 Energy Events:,         505., com/long , 6.95E+11,  1.04E+25, 54.81, 1.04E+25,  6.95E+04, 1.36E-09,
                              403., com/long , 2.67E+11,  2.79E+24, 45.76, 2.79E+24,  2.67E+04, 5.22E-10,
                              698., ast/Cich , 6.24E+11,  2.30E+24, 27.16, 2.30E+24,  1.87E+05, 3.66E-09,
Fatalities for model 3000 (B,F,Ts,G): 2.46E+04, 1.25E+06, 2.40E+02, 5.36E+00,
```

Event Number	Type	Mass (g)	Velocity impact (km/s)	m(H2O) fldPen (km)	Energy Raff (erg)	Ir FluExpl (ug/cm2)	M/A bfatal B (deg)	B? ffatal FS	I? tsfatal	Elev gfatal	Zmin (km)	Vfinal (km/s)	L(burn) (km)	t(burn) (s)	Max mag	Elfinal (deg)	mass% Left	Yield Mton	Xnox (ppm)	Blowoff FB	Crater r(km)	r(4psi) (km)	LS
Offshore runupit dist(km) (cm)																		erg/cm2					
5.	ast/iron	2.84E+09	0.00E+00,	11.43,	1.85E+21,	4.16E-11,	,li	,23.97,	0.00,	9.58,	299.54,	26.39,	-33.97,	27.71,	99.89,	3.15E-02,	0.0,	0.000,	0.13,	1.61,	0.480,		
0.00,		0.00,	3.66,	78130.03,	301.14E+05,	489.91E-02,																	
49.	ast/pall	1.87E+09	0.00E+00,	12.20,	1.39E+21,	1.10E-12,	,if	,41.57,	0.81,	10.56,	179.09,	14.77,	-27.91,	49.62,	99.88,	2.50E-02,	0.0,	0.000,	0.00,	2.17,	2.139,		
0.00,		0.00,	5.24,	1.18,	526.68E+07,	407.31E+00,																	
157.	ast/Och	7.38E+10	0.00E+00,	15.91,	9.34E+22,	7.22E-10,	,if	,78.01,	12.09,	15.87,	110.19,	6.92,	-24.14,	79.76,	99.99,	2.22E+00,	0.0,	0.000,	0.00,	0.00,	0.531,		
0.00,		0.00,	8.96,	211.97E+07,	0.00E+00,	857.00E+01,																	
162.	ast/ach	6.79E+10	0.00E+00,	20.31,	1.40E+23,	1.33E-11,	,if	,32.99,	11.29,	20.09,	202.97,	9.99,	-23.34,	32.17,	99.96,	3.27E+00,	0.0,	0.000,	0.00,	0.00,	0.594,		
0.00,		0.00,	12.71,	358.46E+07,	0.00E+00,	199.08E+00,																	
237.	ast/iron	1.63E+10	0.00E+00,	10.43,	1.06E+22,	4.79E-09,	,if	,31.59,	0.00,	10.71,	229.06,	20.14,	-29.41,	33.20,	99.96,	2.24E-01,	0.0,	0.000,	0.00,	4.34,	0.693,		
0.00,		0.00,	6.21,	1.06E+22,	2.14E-09,	,if	,31.28E-01,																
283.	ast/Och	2.19E+11	0.00E+00,	16.62,	3.02E+23,	2.14E-09,	,if	,56.48,	13.80,	16.60,	127.60,	7.67,	-24.57,	56.40,	99.99,	7.21E+00,	0.0,	0.000,	0.00,	0.00,	0.408,		
0.00,		0.00,	19.99,	528.23E+07,	0.00E+00,	321.27E+00,																	
333.	ast/pell	6.03E+09	0.00E+00,	11.35,	3.88E+21,	3.54E-12,	,oi	,11.98,	0.00,	7.58,	640.36,	57.56,	-31.37,	14.10,	99.81,	4.19E-02,	0.0,	0.000,	0.14,	1.77,	2.707,		
485.92,		378.36,	0.14,	4.03,	85087.69,	188.74E+06,	185.02E-10,																
342.	ast/com/long	5.21E+10	3.90E+10,	0.00,	1.69,	118.26E+07,	0.00E+00,	,ii	,11.14,	39.09,	45.74,	522.00,	11.40,	-24.27,	6.71,	99.50,	1.29E+01,	0.0,	0.000,	0.00,	0.00,	1.311,	
0.00,		0.00,	45.78,	5.46E+23,	1.02E-10,																		
373.	ast/iron	1.94E+09	0.00E+00,	12.44,	1.50E+21,	5.68E-10,	,if	,86.44,	0.64,	11.71,	119.24,	9.61,	-29.43,	90.00,	99.94,	3.17E-02,	0.0,	0.000,	0.00,	2.36,	1.386,		
0.00,		0.00,	6.21,	1.40,	107.75E+08,	854.38E-11,																	
381.	ast/iron	3.44E+10	0.00E+00,	32.29,	1.80E+23,	3.37E-10,	,if	,45.75,	20.96,	32.24,	139.20,	4.31,	-25.89,	45.07,	99.95,	4.28E+00,	0.0,	0.000,	0.00,	0.00,	2.498,		
0.00,		0.00,	0.00,	8.78,	135.76E+07,	762.15E+01,	0.00E+00,																
482.	ast/iron	2.64E+09	0.00E+00,	11.33,	1.69E+21,	1.81E-09,	,if	,86.02,	0.00,	10.60,	119.91,	10.60,	-32.21,	90.00,	99.95,	3.59E-02,	0.0,	0.000,	0.14,	1.68,	1.639,		
0.00,		0.00,	3.83,	81261.58,	237.29E+05,	771.34E-01,																	
640.	ast/CMch	3.74E+11	3.74E+10,	0.00,	30.69,	4.57E+23,	7.60E-10,	,of	,56.19,	21.43,	19.64,	118.90,	6.04,	-24.06,	55.99,	99.99,	1.72E+01,	0.0,	0.000,	0.00,	0.00,	-1.340,	
0.00,		0.00,	30.85,	522.93E+07,	0.00E+00,	682.93E-01,																	
657.	ast/ach	7.69E+09	0.00E+00,	11.30,	4.91E+21,	1.50E-12,	,if	,23.19,	3.29,	9.92,	301.22,	26.79,	-21.71,	25.46,	99.92,	9.03E-02,	0.0,	0.000,	0.00,	0.00,	2.926,		
0.00,		0.00,	0.94,	116.57E+07,	0.00E+00,	232.48E+00,																	
705.	ast/pall	1.84E+09	0.00E+00,	12.79,	1.50E+21,	1.08E-12,	,if	,41.47,	0.56,	11.10,	179.92,	14.16,	-28.96,	49.21,	99.87,	2.70E-02,	0.0,	0.000,	0.00,	2.23,	2.032,		
0.00,		0.00,	6.01,	1.30,	118.39E+08,	336.00E+00,	572.01E-01,																
725.	ast/iron	3.70E+09	0.00E+00,	11.92,	2.63E+21,	1.08E-12,	,if	,60.66,	0.40,	11.25,	136.65,	11.48,	-30.63,	68.17,	99.95,	5.59E-02,	0.0,	0.000,	0.00,	2.59,	0.529,		
0.00,		0.00,	7.71,	1.94,	498.32E+08,	142.18E-01,	0.00E+00,																
751.	ast/CMch	9.71E+10	0.00E+00,	30.69,	4.57E+23,	2.92E-09,	,if	,28.70,	32.80,	30.68,	185.60,	6.05,	-23.72,	27.38,	99.98,	1.09E+01,	0.0,	0.000,	0.00,	0.00,	1.910,		
0.00,		0.00,	14.83,	141.71E+07,	0.00E+00,	561.60E+01,																	
767.	ast/pall	1.71E+09	0.00E+00,	11.51,	1.13E+21,	1.00E-12,	,if	,45.30,	0.32,	10.03,	167.51,	14.63,	-29.55,	54.68,	99.90,	2.05E-02,	0.0,	0.000,	0.00,	1.90,	2.554,		
0.00,		0.00,	5.43,	1.16,	271.82E+08,	809.64E+00,	151.76E+00,																
841.	ast/ach	1.64E+13	0.00E+00,	30.65,	7.68E+25,	3.20E-09,	,oi	,21.23,	20.34,	30.64,	290.00,	9.46,	-26.74,	19.00,	99.99,	1.83E+03,	0.0,	0.000,	3.50,	89.29,	0.775,		
183.66,		209500.80,	1243.08,	0.00,	21004001.00,	556.60E+07,	231.32E+03,																
845.	ast/Och	7.65E+10	0.00E+00,	14.36,	7.89E+22,	7.49E-10,	,if	,28.69,	13.39,	14.27,	226.18,	15.74,	-23.36,	27.79,	99.98,	1.86E+00,	0.0,	0.000,	0.00,	0.00,	1.838,		
0.00,		0.00,	6.29,	144.97E+07,	0.00E+00,	855.65E+00,																	
905.	ast/iron	3.08E+09	0.00E+00,	11.41,	2.01E+21,	9.05E-11,	,oi	,48.72,	0.00,	10.48,	159.19,	13.98,	-31.56,	55.07,	99.94,	4.10E-02,	0.0,	0.000,	0.14,	1.75,	3.078,		
509.75,		356.72,	0.12,	4.00,	84526.71,	324.85E+05,	228.26E-07,	0.00E+00,	149.30E+02,														
955.	ast/Och	2.08E+11	0.00E+00,	20.03,	4.17E+23,	2.03E-09,	,if	,49.06,	15.33,	20.00,	139.20,	6.95,	-25.25,	48.61,	99.99,	9.94E+00,	0.0,	0.000,	0.00,	0.00,	1.707,		
0.00,		0.00,	23.75,	589.89E+07,	0.00E+00,	902.50E+01,																	

Big-3 Masses;
0.00, 0.00, 841.,ast/ach , 1.64E+13, 30.65, 7.68E+25, 1.64E+05, 3.20E-09,
Big-3 Iridium Events;
398.,com/long, 5.62E+11, 60.34, 1.02E+25, 5.62E+04, 1.10E-09,
773.,ast/CMch, 1.18E+12, 29.46, 9.27E+05, 9.27E-09,
237.,ast/CMch, 1.18E+12, 29.46, 1.02E+25, 1.06E+22, 2.45E+05, 4.79E-09,
847.,ast/Och , 4.82E+11, 11.39, 3.09E+23, 2.41E+05, 4.72E-09,
Big-3 Energy Events;
841.,ast/ach , 1.64E+13, 30.65, 7.68E+25, 1.64E+05, 3.20E-09,
398.,com/long, 5.62E+11, 60.34, 1.02E+25, 5.62E+04, 1.10E-09,
773.,ast/CMch, 1.18E+12, 29.46, 5.14E+24, 4.74E+05, 9.27E-09,

Fatalities for model 3001 (B,F,Is,G): 1.79E+03, 2.52E+04, 2.94E+05, 1.24E+03,

Event Number	Type	Mass (g)	m(H2O) impact (km/s)	Velocity (km/s)	Energy (erg)	Ir (ug/cm2)	M/A FluExpl bfatal	B? Raff fFatal	I? FS tsfatal	Elev (deg) gfatal	Zmin (km)	Vfinal (km/s)	L(burn) (km)	t(burn) (s)	Max mag	ELfinal (deg)	mass% left	Yield Mton	Xnox (ppm)	Blowoff FB	Crater r(km)	r(4psi) (km)	LS
Offshore runupht dist(km) (cm)			fldPen R(1psi) (km)			erg/cm2																	
9.	ast/0ch	3.32E-10	0.00E+00,	15.06,	3.76E+22,	3.24E-10,	,1f	,83.53,	10.34,	14.98,	110.18,	7.31,	-23.88,	90.00,	99.98,	8.89E-01,	0.0,0.000,	0.00,	0.00,	1.181,			
	0.00,		0.00,	0.00,	0.00,	2.91,	116.13E+07,	0.00E+00,	,1f	402.41E-01,	0.00E+00,	0.00E+00,											
19.	ast/0ch	1.59E+11,	0.00E+00,	11.93,	1.13E+23,	1.56E-09,	,1f	45.46,	10.72,	11.90,	153.69,	12.86,	-23.39,	45.51,	99.99,	2.69E+00,	0.0,0.000,	0.00,	0.00,	2.569,			
	0.00,		0.00,	0.00,	0.00,	11.30,	326.72E+07,	0.00E+00,	,1f	148.68E+02,	0.00E+00,	0.00E+00,											
80.	ast/ach	2.59E+09,	0.00E+00,	11.35,	1.67E+21,	5.07E-13,	,1f	75.59,	1.47,	10.40,	121.55,	10.74,	-23.19,	90.00,	99.93,	3.35E-02,	0.0,0.000,	0.00,	1.70,	1.227,			
	0.00,		0.00,	0.93,		1.11,	217.01E+07,	307.08E-01,	,1f	653.76E-02,	871.78E-03,												
95.	ast/ach	7.57E+09,	0.00E+00,	11.63,	5.12E+21,	1.48E-12,	,1f	36.31,	3.64,	10.79,	196.98,	16.97,	-21.85,	39.24,	99.94,	1.05E-01,	0.0,0.000,	0.00,	0.00,	2.101,			
	0.00,		0.00,	0.00,		0.83,	110.64E+07,	0.00E+00,	,1f	274.04E-01,	0.00E+00,												
104.	ast/iron,	1.17E+10,	0.00E+00,	13.09,	1.01E+22,	8.04E-09,	,1f	25.88,	2.01,	12.20,	275.80,	21.11,	-28.05,	26.60,	99.94,	2.09E-01,	0.0,0.000,	0.00,	4.16,	1.494,			
	0.00,		0.00,	0.00,		3.51,	724.26E+07,	338.75E+00,	,1f	120.59E+00,	0.00E+00,												
150.	ast/ach	2.48E+10,	0.00E+00,	14.68,	2.67E+22,	4.85E-12,	,1f	39.23,	7.07,	14.31,	179.73,	12.25,	-22.28,	39.79,	99.96,	6.06E-01,	0.0,0.000,	0.00,	0.00,	2.728,			
	0.00,		0.00,	0.00,		4.12,	169.16E+07,	0.00E+00,	,1f	284.27E+01,	0.00E+00,												
388.	com/per	5.32E+11,	2.66E+11,	23.48,	1.47E+24,	2.08E-09,	,1f	19.61,	30.43,	23.48,	281.30,	11.98,	-23.26,	17.55,	99.97,	3.51E+01,	0.0,0.000,	0.00,	0.00,	1.266,			
	0.00,		0.00,	0.00,		44.09,	528.30E+07,	0.00E+00,	,1f	112.64E+02,	0.00E+00,												
396.	com/long,	3.95E+10,	2.96E+10,	46.12,	4.20E+23,	7.73E-11,	,1f	51.01,	36.94,	46.12,	107.30,	2.33,	-24.32,	50.46,	99.89,	1.00E+01,	0.0,0.000,	0.00,	0.00,	0.750,			
	0.00,		0.00,	0.00,		4.14,	102.57E+07,	0.00E+00,	,1f	303.49E-01,	0.00E+00,												
423.	com/long,	8.60E+10,	6.45E+10,	61.35,	1.62E+24,	1.68E-10,	,1f	14.56,	42.45,	61.34,	342.20,	5.58,	-25.62,	11.66,	99.98,	3.86E+01,	0.0,0.000,	0.00,	0.00,	2.324,			
	0.00,		0.00,	0.00,		41.87,	298.53E+07,	0.00E+00,	,1f	116.24E+03,	0.00E+00,												
467.	ast/0ch,	3.35E+10,	0.00E+00,	22.66,	8.60E+22,	3.28E-10,	,1f	59.68,	15.05,	22.59,	121.79,	5.37,	-25.17,	59.69,	99.96,	2.04E+00,	0.0,0.000,	0.00,	0.00,	2.632,			
	0.00,		0.00,	0.00,		5.35,	125.76E+07,	0.00E+00,	,1f	385.24E+01,	0.00E+00,												
503.	ast/0ch,	5.88E+11,	0.00E+00,	12.57,	4.65E+23,	5.75E-09,	,1f	54.81,	8.68,	12.49,	136.28,	10.83,	-26.12,	55.48,	100.00,	1.10E+01,	0.0,0.000,	0.00,	14.14,	1.852,			
	0.00,		0.00,	0.00,		26.69,	203.00E+08,	893.59E+01,	,1f	159.17E+02,	0.00E+00,												
544.	ast/0ch,	1.05E+11,	0.00E+00,	30.67,	4.94E+23,	1.03E-09,	,1f	48.25,	21.02,	30.64,	133.40,	4.35,	-26.20,	47.62,	99.97,	1.18E+01,	0.0,0.000,	0.00,	0.00,	-2.186,			
	0.00,		0.00,	0.00,		24.27,	372.03E+07,	0.00E+00,	,of	603.19E-02,	0.00E+00,												
609.	ast/ach,	1.65E+09,	0.00E+00,	11.49,	1.09E+21,	3.22E-13,	,1f	50.16,	1.46,	10.05,	153.32,	13.43,	-22.84,	61.60,	99.90,	1.98E-02,	0.0,0.000,	0.00,	0.90,	0.765,			
	0.00,		0.00,	0.00,		0.89,	130.60E+07,	297.09E-02,	,1f	581.63E-03,	0.00E+00,												
613.	ast/0ch,	5.45E+10,	0.00E+00,	31.72,	2.74E+23,	5.33E-10,	,1f	66.81,	21.47,	31.70,	107.30,	3.38,	-25.89,	66.66,	99.97,	6.54E+00,	0.0,0.000,	0.00,	0.00,	2.553,			
	0.00,		0.00,	0.00,		0.56,	197.94E+07,	0.00E+00,	,1f	248.62E+02,	0.00E+00,												
623.	ast/iron,	4.12E+10,	0.00E+00,	20.64,	8.79E+22,	1.21E-08,	,1f	48.09,	6.19,	20.41,	153.67,	7.45,	-28.62,	48.19,	99.96,	2.05E+00,	0.0,0.000,	0.00,	5.82,	0.513,			
	0.00,		0.00,	0.00,		14.88,	747.52E+07,	694.27E-01,	,1f	124.55E+00,	0.00E+00,												
630.	ast/0ch,	3.19E+10,	0.00E+00,	23.90,	9.12E+22,	3.12E-10,	,1f	58.42,	16.46,	23.84,	121.79,	5.10,	-25.12,	58.30,	99.96,	2.17E+00,	0.0,0.000,	0.00,	0.00,	3.065,			
	0.00,		0.00,	0.00,		3.93,	111.64E+07,	0.00E+00,	,1f	563.33E+01,	0.00E+00,												
650.	ast/iron,	3.46E+09,	0.00E+00,	19.36,	6.48E+21,	5.07E-12,	,1f	49.69,	3.12,	18.50,	153.57,	7.95,	-28.52,	52.33,	99.87,	1.41E-01,	0.0,0.000,	0.00,	0.75,	1.927,			
	0.00,		0.00,	0.00,		2.21,	201.95E+07,	297.81E-01,	,1f	129.56E+00,	0.00E+00,												
775.	ast/iron,	1.31E+11,	0.00E+00,	14.72,	1.42E+23,	8.95E-09,	,1f	70.98,	1.97,	14.57,	124.67,	8.46,	-30.17,	73.51,	99.99,	3.32E+00,	0.0,0.000,	0.00,	10.52,	0.388,			
	0.00,		0.00,	0.00,		14.99,	118.75E+09,	170.08E+00,	,1f	172.72E+00,	0.00E+00,												
875.	ast/0ch,	4.20E+10,	0.00E+00,	18.76,	7.40E+22,	4.11E-10,	,1f	76.09,	12.97,	18.70,	110.19,	5.87,	-24.53,	77.39,	99.98,	1.76E+00,	0.0,0.000,	0.00,	0.00,	1.665,			
	0.00,		0.00,	0.00,		6.14,	145.74E+07,	0.00E+00,	,1f	547.96E+00,	0.00E+00,												
921.	ast/ach,	6.34E+09,	0.00E+00,	11.70,	4.33E+21,	1.24E-12,	,1f	84.67,	1.33,	10.96,	118.71,	10.17,	-24.17,	90.00,	99.95,	9.10E-02,	0.0,0.000,	0.00,	3.30,	1.211,			
	0.00,		0.00,	0.00,	3.23,		719.54E+07,	111.20E+00,	,1f	273.67E-01,	0.00E+00,	101.89E-01,											

Big-3 Masses:
672.,ast/0ch, 1.32E+12, 1.32E+12, 11.29, 8.44E+23, 6.62E+05, 1.29E-08,
503.,ast/0ch, 5.88E+11, 5.88E+11, 12.57, 4.65E+23, 2.94E+05, 5.75E-09,

Big-3 Iridium Events:
672.,ast/0ch, 1.32E+12, 1.32E+12, 11.29, 8.44E+23, 2.94E+05, 5.75E-09,
388.,com/per, 5.32E+11, 2.66E+11, 23.48, 1.47E+24, 1.06E+05, 2.08E-09,
623.,ast/iron, 4.12E+10, 0.00E+00, 20.64, 8.79E+22, 6.19E+05, 1.21E-08,
775.,ast/iron, 1.31E+11, 0.00E+00, 14.72, 1.42E+23, 4.58E+05, 8.95E-09,

Big-3 Energy Events:
423.,com/long, 8.60E+10, 6.45E+10, 61.35, 1.62E+24, 8.60E+03, 1.68E-10,
388.,com/per, 5.32E+11, 2.66E+11, 23.48, 1.47E+24, 1.06E+05, 2.08E-09,
562.,com/long, 8.81E+10, 46.05, 9.34E+23, 8.81E+03, 1.72E-10,

Fatalities for model 3002 (B,F,Ts,G): 9.69E+05, 1.97E+05, 0.00E+00, 1.11E+01,

Event Number	Type	Mass (g)	Velocity impact (km/s)	Energy (erg)	m(H2O) (ug/cm2)	Ir M/A	B?	I? Elev	FS (deg)	Zmin (km)	Vfinal (km/s)	L(burn) (km)	t(burn) (s)	Max mag	Elfinal (deg)	mass% left	Yield Mton	Xnox (ppm)	Blowoff FB	Crater r(km)	r(4psi) (km)	LS
Offshore dist(km)	runupHt (cm)	fldPen R(1psi) (km)	Raff (km)	FluExpl erg/cm2	bFatal	fFatal	tsFatal gFatal															
15	ccm/per	7.74E+10	24.13	2.25E+23	3.03E-10	,1f	,63.66	26.55	, 24.12	104.40	4.33	-23.31	63.51	99.96	5.38E+00	0.0,0.000	0.00	0.00	1.197
0.00	0.00	0.00	4.73	106.49E+07			0.00E+00	110.69E+00		0.00E+00									
28	ccm/per	1.30E+11	6.51E+10	27.07	4.77E+23	5.10E-10	,1f	,38.53	30.57	, 27.08	145.00	5.35	-22.99	37.66	99.96	1.14E+01	0.0,0.000	0.00	0.00	2.029
0.00	0.00	0.00	17.94	170.28E+07			0.00E+00	108.22E+02		0.00E+00									
30	ast/och	1.59E+11	0.00E+00	20.89	3.47E+23	1.55E-09	,1f	,34.38	16.71	, 20.85	185.59	8.88	-25.08	33.37	99.98	8.25E+00	0.0,0.000	0.00	0.00	0.873
0.00	0.00	0.00	20.66	412.02E+07			0.00E+00	100.16E+01		0.00E+00									
51	ast/och	6.54E+10	0.00E+00	29.02	2.76E+23	6.40E-10	,1f	,19.61	22.06	, 28.95	310.29	10.69	-25.55	17.26	99.93	6.55E+00	0.0,0.000	0.00	0.00	1.948
0.00	0.00	0.00	14.46	187.70E+07			0.00E+00	583.07E+01		0.00E+00									
113	ccm/long	1.78E+11	1.33E+11	53.42	2.54E+24	3.48E-10	,1f	,33.59	39.34	, 53.42	147.90	2.77	-25.63	32.53	99.92	6.06E+01	0.0,0.000	0.00	0.00	1.538
0.00	0.00	0.00	58.19	546.50E+07			0.00E+00	367.40E+02		0.00E+00									
163	ast/och	8.10E+10	0.00E+00	11.32	5.19E+22	7.92E-10	,1f	,16.70	11.23	, 11.11	397.25	35.07	-22.62	15.39	99.98	1.19E+00	0.0,0.000	0.00	0.00	1.311
0.00	0.00	0.00	0.00	4.46	132.27E+07			0.00E+00	128.13E+00		0.00E+00								
165	ast/iron	2.10E+09	0.00E+00	0.00	11.72	1.44E+21	6.16E-10	,pi	,55.76	0.00	10.71	144.51	12.36	-33.98	65.31	99.93	2.92E-02	0.0,0.000	0.00	0.13	0.889
590.20	259.95	0.05	3.57	76333.87	148.68E+06	283.78E-07		0.00E+00	481.79E-01		0.00E+00								
168	ast/CMch	1.21E+11	1.21E+10	22.67	3.12E+23	9.50E-10	,1f	,55.36	24.81	, 22.65	116.00	5.12	-23.72	55.05	99.98	7.44E+00	0.0,0.000	0.00	0.00	1.350
0.00	0.00	0.00	14.39	168.68E+07			0.00E+00	145.81E+01		0.00E+00									
532	ast/iron	1.68E+09	0.00E+00	17.99	2.71E+21	4.92E-13	,1f	,65.06	1.46	, 16.92	130.32	7.27	-29.41	74.69	99.85	5.72E-02	0.0,0.000	0.00	2.52	1.850
0.00	0.00	0.00	1.15	1.69	372.42E+07	282.05E+00		636.48E-01		565.45E-02									
556	ast/Clch	2.46E+11	4.91E+10	25.24	7.83E+23	1.44E-09	,1f	,73.06	25.70	, 25.23	98.60	3.90	-24.62	73.20	99.99	1.87E+01	0.0,0.000	0.00	0.00	1.473
0.00	0.00	0.00	30.89	394.95E+07			0.00E+00	890.35E+01		0.00E+00									
658	ast/och	1.55E+12	0.00E+00	11.34	9.97E+23	1.52E-08	,1f	,22.81	12.29	, 11.33	284.20	25.03	-23.99	21.83	100.00	2.38E+01	0.0,0.000	0.00	0.00	0.911
0.00	0.00	0.00	39.39	219.79E+08	141.35E+01			397.02E+01		0.00E+00									
738	ast/iron	2.65E+09	0.00E+00	11.45	1.74E+21	1.82E-10	,1i	,13.31	0.00	, 0.00	8.06	564.24	50.10	-36.45	15.68	99.80	2.08E-02	0.0,0.000	0.00	0.12	1.011
0.00	0.00	0.00	3.19	69016.24	551.89E+05	126.40E-01		0.00E+00		627.20E-02									
805	ast/Och	1.51E+11	0.00E+00	24.61	4.57E+23	1.48E-09	,1f	,41.52	19.01	, 24.58	153.70	6.24	-25.63	40.72	99.98	1.09E+01	0.0,0.000	0.00	0.00	2.058
0.00	0.00	0.00	23.85	421.02E+07			0.00E+00	204.18E+02		0.00E+00									
829	ast/Clch	4.83E+11	9.67E+10	12.33	3.67E+23	2.84E-09	,1f	,57.04	20.25	, 12.32	118.90	9.63	-22.16	57.35	100.00	8.77E+00	0.0,0.000	0.00	0.00	1.149
0.00	0.00	0.00	19.96	298.28E+07			0.00E+00	176.38E+01		0.00E+00									

Big-3 Masses:;   658..ast/Och , 1.55E+12, 11.34, 9.97E+23, 7.76E+05, 1.52E-08,
                 914..ast/CMch, 4.93E+11, 25.72, 1.63E+24, 1.97E+05, 3.86E-09,
                 829..ast/Clch, 4.83E+11, 12.33, 3.67E+23, 1.45E+05, 2.84E-09,

Big-3 Iridium Events:; 658..ast/Och , 1.55E+12, 11.34, 9.97E+23, 7.76E+05, 1.52E-08,
                        914..ast/CMch, 4.93E+11, 25.72, 1.63E+24, 1.97E+05, 3.86E-09,
                        732..ast/Och , 3.61E+11, 20.89, 7.89E+23, 1.97E+05, 3.53E-09,

Big-3 Energy Events:;   113..ccm/long, 1.78E+11, 53.42, 2.54E+24, 1.78E+04, 3.48E-10,
                        914..ast/CMch, 4.93E+11, 25.72, 1.63E+24, 1.97E+05, 3.86E-09,
                        68..ccm/long, 5.29E+10, 66.52, 1.17E+24, 5.29E+03, 1.03E-10,

Fatalities for model 3003 (B,F,Ts,G): 1.71E+03, 9.12E+04, 4.82E+01, 1.19E+01,

```
Event   Type     Mass         Velocity  m(H2O)    Energy      Ir    M/A     B?   I?   Elev    Zmin    Vfinal   L(burn)  t(burn)  Max    Elfinal  mass%  Yield    Xnox  Blowoff  Crater  r(4psi)  LS
Number           (g)          impact    (erg)     (erg)       Raff  FluExpl R(1psi) B  FS    (deg)   (km)     (km/s)   (km)     (s)    mag     (deg)  left   Mton     (ppm) FB       r(km)   r(km)    (km)
        Offshore runupHt fldPen R(1psi)                       bfatal fFatal   tsFatal gFatal
        dist(km) (cm)  (km)    (km)               erg/cm2

 22.,ast/iron,  2.97E+09, 0.00E+00, 11.29, 1.90E+21, 8.73E-10,   ,li  ,63.54,   0.00, 10.55, 133.37, 11.83, -29.79, 74.85, 99.95, 4.00E-02, 0.0,0.000, 0.14, 1.74, 1.188,
   0.00,    0.00,   0.00,    3.96,83926.61, 504.82E+06, 293.02E-01,    0.00E+00, 145.40E-01,
 96.,ast/ach,  1.37E+11, 0.00E+00, 19.27, 2.54E+23, 2.68E-11,   ,lf ,35.81, 11.12, 19.13, 188.48,  9.78, -23.48, 35.09, 99.98, 5.99E+00, 0.0,0.000, 0.00, 1.80, 1.106,
   0.00,    0.00,   0.00,   18.68,  675.75E+07, 259.23E-01, 139.91E+01,    0.00E+00,
129.,ast/och,  8.04E+10, 0.00E+00, 18.06, 1.31E+23, 7.87E-10,   ,lf ,49.64, 14.34, 18.02, 139.19,  7.70, -24.38, 49.36, 99.99, 3.12E+00, 0.0,0.000, 0.00, 0.00, 2.169,
   0.00,    0.00,   0.00,   10.60,  211.55E+07,    0.00E+00, 520.91E+01,    0.00E+00,
161.,ast/iron, 5.45E+09, 0.00E+00, 12.24, 4.08E+21, 3.73E-11,   ,of ,83.07,  1.17, 11.79, 119.30,  9.75, -28.67, 90.00, 99.97, 9.04E-02, 0.0,0.000, 0.00, 3.36,-1.457,
   0.00,    0.00,   5.15,    2.35,  929.40E+07, 123.73E-02,    0.00E+00,    0.00E+00,
282.,com/long, 3.78E+11, 2.83E+11, 44.77, 3.78E+24, 7.39E-10,   ,of ,37.82, 37.31, 44.77, 136.30,  3.04, -25.37, 36.90, 99.96, 9.04E+01, 0.0,0.000, 0.00, 0.00,-2.402,
   0.00,    0.00,   0.00,   74.16,  906.24E+07,    0.00E+00, 342.63E-01,    0.00E+00,
325.,ast/ach,  4.60E+10, 0.00E+00, 15.13, 5.27E+22, 9.00E-12,   ,lf ,58.16,  6.73, 14.91, 133.36,  8.81, -22.92, 59.53, 99.98, 1.22E+00, 0.0,0.000, 0.00, 0.43, 2.217,
   0.00,    0.00,   0.00,    7.83,  376.40E+07, 189.04E-01, 317.69E+01,    0.00E+00,
348.,ast/meso, 1.94E+09, 0.00E+00, 12.57, 1.54E+21, 7.61E-11,   ,lf ,83.08,  1.16, 11.65, 119.20,  9.51, -27.60, 90.00, 99.92, 3.15E-02, 0.0,0.000, 0.00, 2.12, 2.596,
   0.00,    0.00,   3.69,    1.22,  328.48E+07, 111.28E+01, 185.58E+00,    0.00E+00,
384.,ast/och,  2.99E+10, 0.00E+00, 31.98, 1.53E+23, 2.93E-10,   ,lf ,33.34, 21.28, 31.91, 182.69,  5.71, -25.74, 32.17, 99.93, 3.64E+00, 0.0,0.000, 0.00, 0.00, 0.713,
   0.00,    0.00,   0.00,    5.18,  112.09E+07,    0.00E+00, 435.68E-01,    0.00E+00,
543.,ast/iron, 4.89E+09, 0.00E+00, 16.31, 6.51E+21, 3.55E-10,   ,lf ,50.96,  0.76, 15.39, 153.53,  9.44, -30.96, 55.02, 99.91, 1.38E-01, 0.0,0.000, 0.00, 3.72, 0.424,
   0.00,    0.00,   9.58,    3.03,  331.24E+08, 230.44E-01, 764.53E-02,    0.00E+00, 146.13E-01,
570.,ast/ach,  1.48E+10, 0.00E+00, 11.29, 9.43E+21, 2.69E-12,   ,lf ,16.70,  3.80,  9.79, 422.99, 37.67, -21.64, 17.41, 99.92, 1.69E-01, 0.0,0.000, 0.00, 0.00, 0.262,
   0.00,    0.00,   0.00,    2.12,  163.51E+07,    0.00E+00, 257.96E-02,    0.00E+00,
770.,ast/iron, 2.83E+09, 0.00E+00, 14.66, 3.04E+21, 4.14E-11,   ,of , 9.92,  0.00,  7.28, 954.77, 68.36, -36.26,  8.91, 99.43, 1.80E-02, 0.0,0.000, 0.11, 1.33, 2.176,
 352.48,  342.27,  0.11,    3.04,66083.16, 146.30E+06, 634.56E-11,    0.00E+00, 117.24E+01,    0.00E+00, 5.21E-09,
777.,ast/iron, 1.08E+10, 0.00E+00, 19.76, 2.11E+22, 7.39E-10,   ,lf ,54.08,  5.20, 19.38, 142.04,  7.19, -28.07, 55.24, 99.94, 4.84E-01, 0.0,0.000, 0.00, 0.00, 1.491,
   0.00,    0.00,   0.00,    4.46,  249.91E+07,    0.00E+00, 193.42E+00,    0.00E+00,
881.,ast/och,  9.98E+10, 0.00E+00, 13.40, 8.96E+22, 9.76E-10,   ,lf ,54.51, 11.49, 13.37, 133.39,  9.94, -23.54, 54.75, 99.99, 2.13E+00, 0.0,0.000, 0.00, 0.00, 2.665,
   0.00,    0.00,   0.00,    9.00,  225.04E+07,    0.00E+00, 117.55E+02,    0.00E+00,

Big-3 Masses:
              704.,com/long, 5.11E+12, 36.49, 3.40E+25, 5.11E+05, 9.99E-09,
              486.,com/long, 5.11E+12, 45.42, 8.19E+24, 7.94E+04, 1.55E-09,
              628.,ast/och,  6.27E+11, 17.48, 9.58E+23, 3.14E+05, 6.14E-09,
Big-3 Iridium Events:
              704.,com/long, 5.11E+12, 36.49, 3.40E+25, 5.11E+05, 9.99E-09,
              628.,ast/och,  6.27E+11, 17.48, 9.58E+23, 3.14E+05, 6.14E-09,
              905.,ast/och,  5.33E+11, 16.88, 7.59E+23, 2.66E+05, 5.21E-09,
Big-3 Energy Events:
              704.,com/long, 5.11E+12, 36.49, 3.40E+25, 5.11E+05, 9.99E-09,
              486.,com/long, 7.94E+11, 45.42, 8.19E+24, 7.94E+04, 1.55E-09,
              282.,com/long, 3.78E+11, 44.77, 3.78E+24, 1.55E+04, 7.39E-10,

Fatalities for model 3004 (B,F,Ts,G): 1.21E+03, 2.20E+04, 1.17E+03, 3.51E+02,
```

```
Event  Type     Mass      m(H2O)  Velocity  Energy    Ir        M/A       B?  I? Elev   Zmin  Vfinal  L(burn) t(burn) Max     ElFinal mass%  Yield    Xnox  Blowoff Crater r(4psi) LS
Number          (g)               impact    (erg)     (ug/cm2)  (km)         (deg)  (km)  (km/s)  (km)    (s)     mag     (deg)   left   Mton     (ppm)  FB     r(km)  (km)    (km)
Offshore       runupht   fidPen   R(1psi)   Raff      FluExpl   bfatal   tsfatal  gfatal
dist(km)       (cm)      (km)     (km)                erg/cm2

75.,ast/ach,  1.62E+09, 0.00E+00, 13.48,  1.48E+21, 3.18E-13,    ,lf  ,57.76,   1.80,  12.03, 138.87,  10.36, -23.32,  71.75, 99.87, 2.80E-02, 0.0,0.000,  0.00, 0.51, 1.711,
   0.00,    0.00,    0.00,    0.57, 120.66E+07, 842.68E-02, 529.86E-02,  0.00E+00,  0.00E+00,
140.,ast/Och, 1.65E+09, 0.00E+00, 14.00, 1.61E+21, 1.61E-11,    ,li  , 8.95,   0.00,   3.39,1157.29,  93.16, -30.41,  12.89, 99.12, 2.30E-03, 0.0,0.000, 0.06, 0.67, 2.842,
   0.00,    0.00,    1.53,35625.92, 261.95E+05, 197.26E+00,  0.00E+00, 978.82E-01,
338.,com/long,1.98E+11, 1.48E+11, 31.52, 9.82E+23, 3.87E-10,    ,lf  ,53.00,  32.30,  31.52, 110.20,   3.50, -23.84,  52.50, 99.97, 2.35E+01, 0.0,0.000, 0.00, 0.00, 1.194,
   0.00,    0.00,   33.06, 313.83E+07,  0.00E+00, 537.42E+01,  0.00E+00,  0.00E+00,
354.,ast/Och, 4.67E+13, 0.00E+00, 31.72, 2.35E+26, 4.57E-07,    ,oi  ,27.58,  19.48,  31.72, 223.30,   7.04, -30.67,  25.95,100.00, 5.61E+03, 0.0,0.000, 4.90,121.09, 2.445,
2427.48,27724.03,  83.30,  0.00,32937611.00, 804.55E+05, 349.47E-07,  0.00E+00, 112.61E+05,  0.00E+00,
385.,ast/ach, 8.75E+10, 0.00E+00, 28.40, 3.52E+23, 1.71E-11,    ,lf  ,61.51,  13.14,  28.30, 121.79,   4.29, -24.95,  61.56, 99.96, 8.35E+00, 0.0,0.000, 0.00, 0.00, 1.611,
   0.00,    0.00,    0.00,   22.05, 675.11E+07,  0.00E+00, 624.29E+01,  0.00E+00,  0.00E+00,
494.,ast/iron,1.18E+10, 0.00E+00, 11.29, 7.51E+21, 1.73E-10,    ,oi  ,17.94,   0.00,   9.69, 404.71,  36.04, -34.25,  19.20, 99.93, 1.33E-01, 0.0,0.000, 0.20, 2.60, 2.161,
 583.27, 561.67,   0.31,    5.92,%120373.90, 352.10E+05, 414.89E-08,  0.00E+00, 517.27E+01,  0.00E+00,
544.,ast/pall, 2.32E+09, 0.00E+00, 13.93, 2.25E+21, 1.36E-12,   ,lf  ,49.94,   1.89,  12.78, 153.97,  11.09, -27.10,  56.32, 99.89, 4.53E-02, 0.0,0.000, 0.00, 1.26, 1.237,
   0.00,    0.00,    0.00,    1.16, 177.09E+07, 172.30E-01, 731.39E-02,  0.00E+00,  0.00E+00,
571.,ast/pall, 3.77E+09, 0.00E+00, 20.44, 7.88E+21, 2.21E-12,   ,lf  ,75.81,   4.55,  19.77, 118.80,   5.82, -27.43,  90.00, 99.88, 1.76E-01, 0.0,0.000, 0.00, 0.00, 0.655,
   0.00,    0.00,    0.00,    1.38, 118.68E+07,  0.00E+00, 268.97E-02,  0.00E+00,  0.00E+00,
601.,ast/ach, 2.95E+10, 0.00E+00, 29.17, 1.25E+23, 2.88E-10,    ,lf  ,82.92,  19.24,  29.14, 101.50,   3.48, -25.52,  83.97, 99.96, 2.99E+00, 0.0,0.000, 0.00, 0.00, 1.724,
   0.00,    0.00,    0.00,    4.78, 112.61E+07,  0.00E+00, 380.98E+00,  0.00E+00,  0.00E+00,
669.,ast/iron, 8.77E+21, 0.00E+00, 11.47, 5.77E+21, 2.57E-09,   ,lf  ,63.14,   0.63,  11.01, 137.20,  11.97, -29.93,  65.36, 99.97, 1.27E-01, 0.0,0.000, 3.52, 3.52, 2.300,
   0.00,    0.00,    9.89,    2.91, 444.92E+08, 155.13E+01, 531.96E+00,  0.00E+00, 117.17E+01,
679.,ast/Och, 5.75E+10, 0.00E+00, 11.40, 3.74E+22, 5.62E-10,    ,lf  ,42.09,   9.52,  11.31, 165.28,  14.47, -22.94,  42.36, 99.99, 8.79E-01, 0.0,0.000, 0.00, 0.00, 1.792,
   0.00,    0.00,    0.00,    3.96, 135.31E+07,  0.00E+00, 304.89E+00,  0.00E+00,  0.00E+00,
747.,com/long, 4.04E+10, 3.03E+10, 67.13, 9.09E+23, 7.90E-11,   ,lf  ,39.53,  41.35,  67.12, 124.70,   1.86, -25.75,  38.70, 99.81, 2.17E+01, 0.0,0.000, 0.00, 0.00, 1.360,
   0.00,    0.00,    0.00,   25.40, 176.98E+07,  0.00E+00, 464.08E+01,  0.00E+00,  0.00E+00,
818.,ast/Och, 4.16E+10, 0.00E+00, 13.05, 3.54E+22, 4.07E-10,    ,lf  ,73.58,   8.54,  12.96, 115.98,   8.88, -23.78,  76.22, 99.99, 8.35E-01, 0.0,0.000, 0.00, 0.00, 1.810,
   0.00,    0.00,    0.00,    4.62, 159.65E+07,  0.00E+00, 432.75E+00,  0.00E+00,  0.00E+00,
859.,ast/ach, 2.36E+11, 0.00E+00, 21.81, 5.62E+23, 4.62E-11,    ,lf  ,79.10,  11.74,  21.76, 110.19,   5.05, -24.22,  80.38, 99.99, 1.34E+01, 0.0,0.000, 0.00,10.42, 0.813,
   0.00,    0.00,    0.00,   29.08, 135.29E+08, 443.30E+00, 172.64E+01,  0.00E+00,  0.00E+00,
890.,ast/ach, 8.17E+09, 0.00E+00, 11.60, 5.49E+21, 1.60E-12,    ,li  ,10.59,   0.00,   6.48, 757.16,  67.39, -31.07,  13.30, 99.77, 4.16E-02, 0.0,0.000, 0.14, 1.76, 1.336,
   0.00,    0.00,    0.00,    4.02,84924.59, 619.11E+05, 423.33E-01,  0.00E+00, 210.06E-01,

Big-3 Masses:        354.,ast/Och , 4.67E+13, 31.72, 2.35E+26, 2.33E+07, 4.57E-07,
                     439.,ast/Cich, 2.91E+11, 25.24, 9.27E+23, 8.74E+04, 1.71E-09,
                     859.,ast/ach , 2.36E+11, 21.81, 5.62E+23, 2.36E+03, 4.62E-11,
Big-3 Iridium Events:354.,ast/Och , 4.67E+13, 31.72, 2.35E+26, 2.33E+07, 4.57E-07,
                     669.,ast/iron, 8.77E+09, 11.47, 5.77E+21, 1.32E+05, 2.57E-09,
                     439.,ast/Cich, 2.91E+11, 25.24, 9.27E+23, 8.74E+04, 1.71E-09,
Big-3 Energy Events: 354.,ast/Och , 4.67E+13, 31.72, 2.35E+26, 2.33E+07, 4.57E-07,
                     782.,com/long, 8.81E+10, 63.77, 1.79E+24, 8.81E+03, 1.72E-10,
                     338.,com/long, 1.98E+11, 31.52, 9.82E+23, 1.98E+04, 3.87E-10,

Fatalities for model 3005 (B,F,Ts,G): 2.26E+03, 1.97E+04, 1.13E+07, 1.29E+03,
```

```
Event  Type         Mass      m(H2O)   Velocity  Energy     Ir  M/A      B?    I?  Elev    Zmin  Vfinal  L(burn)  t(burn)  Max     Elfinal  mass%   Yield     Xnox   Blowoff  Crater  r(4psi)  LS
Number              (g)       impact   (km/s)    (erg)      Raff FluExpl bFatal fFatal (deg) (km)  (km/s)  (km)     (s)      mag     (deg)    left    Mton      (ppm)  FB       r(km)   (km)     (km)
       Offshore runupHt fldPen R(1psi)
       dist(km) (cm)       (km)      erg/cm2                           tsFatal gFatal
138.,ast/iron, 3.56E+09, 0.00E+00, 22.45, 8.97E+21, 2.44E-10,  ,1f ,55.66,  5.29,  21.83, 139.12,   6.20, -28.29,  57.64,  99.83, 2.02E-01, 0.0,0.000,  0.00,  0.00, 1.398,
    0.00,  0.00,  0.00,  0.00,   0.34, 100.86E+07,          0.00E+00, 930.00E-03,  0.00E+00,  0.00E+00,
192.,ast/Och , 9.70E+10, 0.00E+00, 11.68, 6.62E+22, 9.49E-10,  ,1f ,25.40, 11.31,  11.58, 258.07,  22.06, -22.88,  24.72,  99.99, 1.55E+00, 0.0,0.000,  0.00,  0.00, 0.294,
    0.00,  0.00,  0.00,  0.00,   6.61, 169.64E+07,          0.00E+00, 269.89E-01,  0.00E+00,  0.00E+00,
273.,ast/ach , 2.53E+10, 0.00E+00, 13.37, 2.26E+22, 4.95E-12,  ,1f ,55.80,  4.85,  13.02, 139.12,  10.41, -22.73,  58.20,  99.97, 5.13E-01, 0.0,0.000,  0.00,  0.97, 3.304,
    0.00,  0.00,  0.00,  0.00,   4.85, 303.77E+07, 119.23E+01, 148.76E+02,  0.00E+00,  0.00E+00,
377.,ast/iron, 2.82E+09, 0.00E+00, 11.68, 1.92E+21, 1.93E-10,  ,1f ,48.23,  0.10,  10.77, 160.29,  13.75, -33.15,  54.09,  99.94, 3.90E-02, 0.0,0.000,  0.00,  1.93, 1.698,
    0.00,  0.00,  0.00,  0.00,   5.75,   1.63, 496.75E+09, 116.64E+00, 417.22E-01,  0.00E+00, 989.11E-01,
380.,ast/iron, 1.96E+10, 0.00E+00, 14.44, 2.04E+22, 2.87E-11,  ,1f ,46.32,  2.57,  14.03, 162.82,  11.28, -28.51,  47.68,  99.97, 4.61E-01, 0.0,0.000,  0.00,  5.44, 2.955,
    0.00,  0.00,  0.00,  0.00,   5.31, 970.40E+07, 167.78E+02, 799.49E-01,  0.00E+00,  0.00E+00,
422.,com/per , 8.45E+10, 4.23E+10, 26.25, 2.91E+23, 3.31E-10,  ,1f ,76.31, 27.02,  26.25,  95.70,   3.64, -23.65,  76.48,  99.96, 6.95E+00, 0.0,0.000,  0.00,  0.00, 2.184,
    0.00,  0.00,  0.00,  0.00,  10.84, 132.88E+07,          0.00E+00, 564.49E+01,  0.00E+00,  0.00E+00,
477.,ast/iron, 1.04E+10, 0.00E+00, 23.07, 2.78E+22, 3.06E-09,  ,1f ,34.25,  7.56,  22.59, 202.94,   8.80, -28.13,  33.80,  99.87, 6.35E-01, 0.0,0.000,  0.00,  0.00, 0.602,
    0.00,  0.00,  0.00,  0.00,   3.93, 155.23E+07,          0.00E+00, 194.27E-01,  0.00E+00,  0.00E+00,
613.,ast/Och , 2.99E+11, 0.00E+00, 13.52, 2.73E+23, 2.93E-09,  ,1f ,76.50,  9.87,  13.50, 113.10,   8.36, -24.63,  78.14, 100.00, 6.52E+00, 0.0,0.000,  0.00,  6.63,-1.278,
    0.00,  0.00,  0.00,  0.00,  19.95, 933.80E+07, 727.85E-02, 329.28E-01,  0.00E+00,  0.00E+00,
702.,com/long, 8.72E+10, 6.54E+10, 47.53, 9.85E+23, 1.71E-10,  ,1f ,37.00, 37.17,  47.53, 139.20,   2.93, -24.92,  36.05,  99.90, 2.35E+01, 0.0,0.000,  0.00,  0.00, 2.374,
    0.00,  0.00,  0.00,  0.00,  30.51, 237.50E+07,          0.00E+00, 692.00E+02,  0.00E+00,  0.00E+00,
708.,com/long, 9.40E+10, 7.05E+10, 45.50, 9.73E+23, 1.84E-10,  ,1f ,55.64, 36.45,  45.50, 101.50,   2.23, -24.77,  55.18,  99.93, 2.32E+01, 0.0,0.000,  0.00,  0.00, 0.923,
    0.00,  0.00,  0.00,  0.00,  30.62, 244.02E+07,          0.00E+00, 246.72E+01,  0.00E+00,  0.00E+00,
732.,ast/Och , 5.48E+10, 0.00E+00, 32.15, 2.83E+23, 5.36E-10,  ,1f ,25.15, 22.35,  32.08, 237.79,   7.40, -25.93,  23.41,  99.93, 6.74E+00, 0.0,0.000,  0.00,  0.00, 1.236,
    0.00,  0.00,  0.00,  0.00,  14.70, 188.29E+07, 116.88E+01,         0.00E+00,  0.00E+00,
744.,ast/Och , 9.66E+10, 0.00E+00, 30.02, 4.35E+23, 9.45E-10,  ,1f ,44.71, 20.75,  29.99, 142.10,   4.73, -26.10,  43.98,  99.97, 1.04E+01, 0.0,0.000,  0.00,  0.00, 2.947,
    0.00,  0.00,  0.00,  0.00,  22.34, 336.49E+07,          0.00E+00, 138.66E+03,  0.00E+00,  0.00E+00,
862.,ast/iron, 2.39E+09, 0.00E+00, 11.43, 1.56E+21, 7.00E-10,  ,oi ,43.72,  0.00,  10.34, 173.20,  15.20, -31.04,  49.75,  99.93, 3.09E-02, 0.0,0.000,  0.13,  1.60, 1.343,
    465.22, 339.37,  0.11,  3.64, 77658.84, 163.64E+05, 562.22E-11,     0.00E+00, 222.81E+00,  0.00E+00,
951.,com/per , 9.29E+11, 4.65E+11, 33.48, 5.21E+24, 3.64E-09,  ,1f ,39.90, 29.71,  33.48, 142.10,   4.24, -25.95,  39.02,  99.97, 1.24E+02, 0.0,0.000,  0.00,  7.71, 1.752,
    0.00,  0.00,  0.00,  0.00,  89.87, 196.68E+08, 210.94E+01, 143.27E+03,  0.00E+00,  0.00E+00,
Big-3 Masses:;                                   951.,com/per , 9.29E+11, 7.97E+11, 33.48, 5.21E+24, 1.86E+05, 3.99E+05, 3.64E-09,
                                                 548.,ast/Och , 9.29E+11, 7.97E+11, 19.54, 1.52E+24, 1.52E+05, 3.99E+05, 7.80E-09,
                                                 548.,ast/Och , 3.96E+11, 3.96E+11, 30.47, 1.84E+24, 1.84E+05, 1.98E+05, 7.80E-09,
                                                 338.,ast/Och , 3.96E+11, 3.96E+11, 30.47, 1.84E+24, 1.84E+05, 1.98E+05, 3.87E-09,
Big-3 Iridium Events:;                           548.,ast/Och , 7.97E+11, 7.97E+11, 19.54, 1.52E+24, 1.52E+05, 3.99E+05, 7.80E-09,
                                                 338.,ast/Och , 3.96E+11, 3.96E+11, 30.47, 1.84E+24, 1.84E+05, 1.98E+05, 3.87E-09,
                                                 951.,com/per , 9.29E+11, 9.29E+11, 33.48, 5.21E+24, 1.86E+05, 3.99E+05, 3.64E-09,
Big-3 Energy Events:;                            951.,com/per , 9.29E+11, 9.29E+11, 33.48, 5.21E+24, 1.86E+05, 3.99E+05, 3.64E-09,
                                                 209.,com/long, 1.99E+11, 57.47, 3.28E+24, 1.86E+05, 1.99E+04, 3.89E-10,
                                                 338.,ast/Och , 3.96E+11, 3.96E+11, 30.47, 1.84E+24, 1.84E+05, 1.98E+05, 3.87E-09,
Fatalities for model 3006 (B,F,Ts,G):  2.02E+04, 3.83E+05, 2.23E+02, 9.89E+01,
```

```
Event Type   Mass      m(H2O)  Velocity Energy     Ir      M/A     B? I?  Elev  F$  B  Zmin  Vfinal L(burn) t(burn) Max    ELfinal mass% Yield   Xnox Blowoff Crater r(4psi) LS
Number       (g)       impact  (km/s)   (erg)      (ug/cm2)                (deg)            (km)  (km/s)  (km)    (s)    meg    (deg)   left  Mton    (ppm) FB      r(km)  (km)
Offshore runupHt fldPen R(1psi) Raff FluExpl  bfatal ffatal tsFatal gFatal
dist(km) (cm)    (km)   (km)    (km) erg/cm2
26.,ast/meso, 2.51E+09, 0.00E+00, 11.73, 1.73E+21, 9.82E-11,   ,oi , 12.21,   ,  , 0.00,   7.02, 626.81,  54.87, -31.79, 16.03, 99.72, 1.50E-02, 0.0,0,0.000, 0.10, 1.25, 0.424,
596.86, 184.35,    0.01,    2.86,62531.21, 308.19E+05, 966.69E-07, 0.00E+00, 353.83E-02, 0.00E+00,
42.,ast/Och , 3.10E+10, 0.00E+00, 26.16, 1.06E+23, 3.03E-10,   ,lf ,42.99, 17.99, 26.09, 150.79,   5.76, -25.30, 42.31, 99.95, 2.52E+00, 0.0,0,0.000, 0.00, 0.00, 1.859,
   0.00,   0.00,    0.00,    3.69, 108.57E+07, 0.00E+00, 308.38E+00, 0.00E+00, 0.00E+00, 0.00E+00,
43.,ast/Och , 3.61E+11, 0.00E+00, 17.31, 5.41E+23, 3.53E-09,   ,lf ,43.47, 14.93, 17.30, 153.70,   8.87, -24.84, 42.91, 99.99, 1.29E+01, 0.0,0,0.000, 0.00, 0.28, 0.586,
   0.00,   0.00,    0.00,   27.79, 807.03E+07, 0.00E+00, 935.31E+00, 0.00E+00, 0.00E+00, 0.00E+00,
76.,ast/ach , 2.75E+11, 0.00E+00, 11.35, 1.77E+23, 5.37E-11,   ,lf ,64.92,  4.21, 11.25, 127.57,  11.23, -23.53, 66.93, 99.99, 4.15E+00, 0.0,0,0.000, 0.00,11.99, 1.419,
   0.00,   0.00,    0.00,   16.59, 326.58E+08, 237.29E+01, 226.98E+01, 0.00E+00, 0.00E+00, 0.00E+00,
88.,ast/Och , 1.23E+12, 0.00E+00, 20.70, 2.63E+24, 1.20E-08,   ,lf ,12.82, 20.06, 20.65, 530.69,  25.63, -25.53,  8.92, 99.98, 6.26E+01, 0.0,0,0.000, 0.00,16.40, 2.680,
   0.00,   0.00,    0.00,   63.90, 217.13E+08, 808.84E+02, 613.56E+03, 0.00E+00, 0.00E+00, 0.00E+00,
99.,ast/ach , 1.94E+11, 0.00E+00, 16.52, 2.65E+23, 3.80E-11,   ,of ,33.83, 10.12, 16.39, 200.08,  12.10, -23.10, 33.13, 99.98, 6.23E+00, 0.0,0,0.000, 0.00, 5.43,-2.015,
   0.00,   0.00,    0.00,   19.39, 848.70E+07, 893.17E-03, 570.56E-02, 0.00E+00, 0.00E+00, 0.00E+00,
143.,ast/CMch, 1.12E+11, 1.12E+10, 24.30, 3.32E+23, 8.80E-10,   ,lf ,67.86, 26.05, 24.29, 101.50,   4.18, -23.47, 67.78, 99.99, 7.93E+00, 0.0,0,0.000, 0.00, 0.00, 1.992,
   0.00,   0.00,    0.00,   14.48, 163.05E+07, 0.00E+00, 647.17E+01, 0.00E+00, 0.00E+00, 0.00E+00,
252.,ast/Och , 8.55E+10, 0.00E+00, 32.30, 4.46E+23, 8.37E-10,   ,lf ,47.69, 21.92, 32.28, 133.40,   4.13, -26.20, 47.04, 99.97, 1.06E+01, 0.0,0,0.000, 0.00, 0.00, 0.861,
   0.00,   0.00,    0.00,   22.20, 309.34E+07, 0.00E+00, 112.46E+01, 0.00E+00, 0.00E+00, 0.00E+00,
259.,ast/iron, 2.88E+09, 0.00E+00, 11.55, 1.92E+21, 4.22E-11,   ,li ,70.58,  0.00, 10.84, 126.56,  10.97, -29.84, 90.00, 99.95, 4.10E-02, 0.0,0,0.000, 0.14, 1.75, 2.366,
   0.00,   0.00,    4.00,84517.59, 264.78E+06, 448.28E+00, 222.44E+00,
319.,ast/ach , 1.10E+10, 0.00E+00, 12.77, 8.93E+21, 2.14E-12,   ,lf ,49.03,  4.00, 12.19, 153.57,  12.04, -22.44, 52.18, 99.95, 1.94E-01, 0.0,0,0.000, 0.00, 0.00, 0.454,
   0.00,   0.00,    0.00,    2.34, 169.69E+07, 0.00E+00, 487.69E-02, 0.00E+00, 0.00E+00, 0.00E+00,
333.,ast/Cich, 4.26E+11, 8.53E+10, 11.91, 3.02E+23, 2.50E-09,   ,lf ,17.84, 28.55, 11.91, 310.30,  26.02, -20.74, 16.49, 99.99, 7.22E+00, 0.0,0,0.000, 0.00, 0.00, 1.597,
   0.00,   0.00,    0.00,    9.73, 123.72E+07, 0.00E+00, 117.72E+01, 0.00E+00, 0.00E+00, 0.00E+00,
378.,ast/Och , 3.90E+10, 0.00E+00, 32.34, 2.04E+23, 3.82E-10,   ,lf ,42.48, 20.98, 32.29, 147.90,   4.57, -25.97, 41.68, 99.95, 4.86E+00, 0.0,0,0.000, 0.00, 0.00, 1.847,
   0.00,   0.00,    0.00,   10.80, 154.07E+07, 0.00E+00, 257.64E+01, 0.00E+00, 0.00E+00, 0.00E+00,
384.,ast/ach , 1.09E+10, 0.00E+00, 11.70, 7.45E+21, 2.13E-12,   ,lf ,32.91,  4.16, 10.94, 214.40,  18.35, -21.79, 34.94, 99.95, 1.55E-01, 0.0,0,0.000, 0.00, 0.00, 1.975,
   0.00,   0.00,    0.00,    1.46, 125.13E+07, 0.00E+00, 633.17E-01, 0.00E+00, 0.00E+00, 0.00E+00,
432.,ast/iron, 2.30E+09, 0.00E+00, 11.44, 1.51E+21, 3.38E-12,   ,li ,71.67,  0.00, 10.64, 125.69,  11.01, -32.23, 90.00, 99.95, 3.16E-02, 0.0,0,0.000, 0.13, 1.61, 1.460,
   0.00,   0.00,    0.00,    3.66,78184.48, 616.75E+05, 468.41E-01, 0.00E+00, 0.00E+00, 232.43E-01,
```

```
503.,ast/meso, 1.71E+09, 0.00E+00, 11.53, 1.14E+21, 6.68E-11,   ,lf ,56.32,  0.37, 10.33, 142.83, 12.44, -29.42, 68.96, 99.91, 2.17E-02, 0.0,0.000, 0.00, 1.97, 1.361,
     0.00,   0.00,  0.00,  5.60,  1.19, 225.13E+08, 560.47E-01, 102.60E-01,  0.00E+00, 432.90E-01,  0.00E+00,
530.,ast/pall, 1.93E-09, 0.00E+00, 16.52, 2.63E+21, 1.13E-12,   ,lf ,62.00,  1.97, 15.33, 133.18,  8.09, -27.78, 72.22, 99.85, 5.41E-02, 0.0,0.000, 0.00, 1.43, 1.385,
     0.00,   0.00,  0.00,  0.00,  1.34, 194.90E+07, 312.76E-01, 137.26E-01,  0.00E+00,  0.00E+00,  0.00E+00,
570.,ast/Och , 1.13E+11, 0.00E+00, 29.55, 4.95E+23, 1.11E-09,   ,lf ,84.86, 18.88, 29.53, 101.50,  3.43, -26.40, 86.03, 99.98, 1.18E-01, 0.0,0.000, 0.00, 0.00, 1.038,
     0.00,   0.00,  0.00,  0.00, 25.17, 462.83E+07,  0.00E+00, 217.09E+01,  0.00E+00,  0.00E+00,  0.00E+00,
605.,ast/Och , 8.10E+10, 0.00E+00, 21.68, 1.90E+23, 7.92E-10,   ,lf ,30.41, 17.88, 21.62, 205.89,  9.49, -24.68, 29.18, 99.97, 4.52E+00, 0.0,0.000, 0.00, 0.00, 1.055,
     0.00,   0.00,  0.00,  0.00, 12.36, 197.62E+07,  0.00E+00, 544.82E+00,  0.00E+00,  0.00E+00,  0.00E+00,
609.,ast/ach , 1.42E+10, 0.00E+00, 11.33, 9.14E+21, 2.79E-12,   ,lf ,23.98,  4.19, 10.38, 289.75, 25.66, -21.66, 25.26, 99.94, 1.83E-01, 0.0,0.000, 0.00, 0.00, 1.120,
     0.00,   0.00,  0.00,  0.00,  1.98, 145.74E+07,  0.00E+00, 162.82E-01,  0.00E+00,  0.00E+00,  0.00E+00,
706.,ast/iron, 1.70E+09, 0.00E+00, 13.37, 1.52E+21, 4.99E-13,   ,lf ,48.40,  1.36, 12.33, 158.42, 11.89, -27.96, 53.91, 99.90, 3.09E-02, 0.0,0.000, 0.00, 1.79, 2.148,
     0.00,   0.00,  0.00,  1.97,  1.10, 233.58E+07, 283.28E+00, 533.45E-01,  0.00E+00, 328.75E-01,  0.00E+00,
745.,ast/iron, 1.18E+10, 0.00E+00, 25.14, 3.75E+22, 8.07E-10,   ,lf ,56.92,  8.51, 24.88, 133.37,  5.31, -28.30, 57.38, 99.91, 8.72E-01, 0.0,0.000, 0.00, 0.00, 0.999,
     0.00,   0.00,  0.00,  0.00,  4.91, 168.11E+07,  0.00E+00, 757.62E-01,  0.00E+00,  0.00E+00,  0.00E+00,
898.,ast/iron, 1.33E+10, 0.00E+00, 13.05, 1.13E+22, 3.90E-10,   ,of ,32.23,  1.93, 12.38, 223.65, 17.16, -28.28, 33.33, 99.95, 2.43E-01, 0.0,0.000, 0.00, 4.52, -1.18,
     0.00,   0.00,  0.00,  0.00,  3.85, 908.97E+07, 420.15E-02, 152.15E-02,  0.00E+00,  0.00E+00,  0.00E+00,
990.,com/per , 2.35E+11, 1.18E+11, 33.16, 1.29E+24, 9.20E-10,   ,of ,49.17, 30.47, 33.16, 118.90,  3.58, -24.75, 48.57, 99.96, 3.09E+01, 0.0,0.000, 0.00, 0.00, -3.592,
     0.00,   0.00,  0.00,  0.00, 40.70, 464.23E+07,  0.00E+00, 666.43E-03,  0.00E+00,  0.00E+00,  0.00E+00,
997.,ast/iron, 3.17E+09, 0.00E+00, 11.51, 2.10E+21, 9.30E-11,   ,oi ,52.73,  0.00, 10.66, 150.23, 13.08, -30.15, 59.50, 99.95, 4.36E-02, 0.0,0.000, 0.14, 1.79, 1.334,
   330.29, 567.88,  0.31,  4.08,86114.74, 604.87E+06, 652.64E+08,  0.00E+00, 445.71E+00,  0.00E+00,
Big-3 Masses:,             88.,ast/Och , 1.23E+12, 20.70, 2.63E+24, 6.14E+05, 1.20E-08,
                          960.,ast/Och , 5.55E+11, 20.68, 1.19E+24, 2.77E+05, 5.43E-09,
Big-3 Iridium Events:,    333.,ast/CIch, 4.26E+11, 11.91, 3.02E+23, 1.28E+05, 2.50E-09,
                           88.,ast/Och , 1.23E+12, 20.70, 2.63E+24, 6.14E+05, 1.20E-08,
                          542.,ast/iron, 2.93E+10, 11.34, 1.89E+22, 4.40E+05, 8.61E-09,
Big-3 Energy Events:,     960.,ast/Och , 5.55E+11, 20.68, 1.19E+24, 2.77E+05, 5.43E-09,
                           88.,ast/Och , 1.23E+12, 20.70, 2.63E+24, 6.14E+05, 1.20E-08,
                          990.,com/per , 2.35E+11, 33.16, 1.29E+24, 4.70E+04, 9.20E-10,
                          798.,ast/Och , 3.42E+11, 27.08, 1.26E+24, 1.71E+05, 3.35E-09,
Fatalities for model 3007 (B,F,Ts,G): 8.41E+04, 6.31E+05, 4.49E+02, 3.22E+02,
```

```
Event Type      Mass       Velocity  m(H2O)    Energy    Ir   M/A     B?   I?  Elev  Zmin  Vfinal L(burn) t(burn)  Max    Elfinal mass%  Yield    Xnox Blowoff Crater r(4psi) LS
Number          (g)        impact    fld-Pen   (erg)         (ug/cm2) B         FS             (km/s)  (km)    (s)    mag    (deg)   left   Mton     (ppm) FB      r(km)   (km)
Offshore runupHt fld-Pen R(1psi)  Raff   FluExpl bFatal fFatal   tsFatal gFatal
dist(km) (cm)    (km)     (km)                 erg/cm2
112., com/per  , 4.18E+11, 2.09E+11, 30.66, 1.96E+24, 1.63E-09,  ,If , 36.73, 30.87,  30.66, 150.80,   4.92, -24.80, 35.76, 99.97, 4.69E+01,  0.0,0.000,  0.00, 0.00, 1.346,
  0.00, 0.00,    0.00,    0.00,    52.35, 686.89E+07,          0.00E+00, 190.94E+02, 0.00E+00, 0.00E+00,
209., ast/CMch, 1.72E+11, 1.72E+10, 16.08, 2.22E+23, 1.34E-09,  ,If , 32.05, 24.38,  16.05, 182.70,  11.35, -22.28, 31.18, 99.99, 5.29E+00,  0.0,0.000,  0.00, 0.00, 1.012,
  0.00, 0.00,    0.00,    0.00,     8.40, 124.22E+07,          0.00E+00, 228.02E+00, 0.00E+00, 0.00E+00,
218., ast/ach , 1.52E+10, 0.00E+00, 11.40, 9.87E+21, 2.97E-12,  ,If , 57.79,  1.91,  10.83, 139.05,  12.21, -23.73, 63.53, 99.97, 2.13E-01,  0.0,0.000,  0.00, 4.28, 1.990,
  0.00, 0.00,    0.00,    0.00,     3.57, 811.01E+07, 112.33E+01, 0.00E+00, 392.02E+00, 0.00E+00, 0.00E+00,
236., ast/Och , 1.40E+11, 0.00E+00, 24.44, 4.18E+23, 1.37E-09,  ,If , 15.85, 20.39,  24.35, 403.09,  16.49, -25.37, 12.82, 99.95, 9.91E+00,  0.0,0.000,  0.00, 0.00, 1.627,
  0.00, 0.00,    0.00,    0.00,    21.78, 332.77E+07,          0.00E+00, 631.42E+01, 0.00E+00, 0.00E+00,
249., ast/long, 3.91E+10, 2.93E+10, 46.61, 4.24E+23, 7.64E-11,  ,If , 67.66, 36.93,  46.61,  89.90,   1.93, -24.33, 67.41, 99.91, 1.01E+01,  0.0,0.000,  0.00, 0.00, 1.675,
  0.00, 0.00,    0.00,    0.00,     4.96, 103.68E+07,          0.00E+00, 365.84E+00, 0.00E+00, 0.00E+00,
309., ast/Och , 7.89E+10, 0.00E+00, 19.97, 1.57E+23, 7.71E-10,  ,If , 53.09, 15.98,  19.94, 130.50,   6.53, -24.55, 52.81, 99.99, 3.75E+00,  0.0,0.000,  0.00, 0.00, 0.444,
  0.00, 0.00,    0.00,    0.00,    11.45, 204.72E+07,          0.00E+00, 114.48E+00, 0.00E+00, 0.00E+00,
368., ast/iron, 4.31E+11, 0.00E+00, 17.98, 6.96E+23, 2.95E-09,  ,If , 61.04,  5.95,  17.92, 130.49,   7.25, -29.23, 61.26, 99.99, 1.65E+01,  0.0,0.000,  0.00, 19.11, 2.189,
  0.00, 0.00,    0.00,    0.00,    33.37, 651.57E+08, 354.77E+02, 541.02E+02, 0.00E+00, 0.00E+00,
391., com/long, 8.42E+10, 6.32E+10, 31.98, 4.31E+23, 1.65E-10,  ,If , 18.70, 34.19,  31.96, 284.20,   8.89, -23.29, 16.47, 99.87, 1.03E+01,  0.0,0.000,  0.00, 0.00, 2.005,
  0.00, 0.00,    0.00,    0.00,    11.37, 122.56E+07,          0.00E+00, 410.56E+01, 0.00E+00, 0.00E+00,
401., ast/ach , 6.08E+09, 0.00E+00, 12.14, 4.48E+21, 1.19E-12,  ,If , 59.41,  2.33,  11.38, 136.11,  11.24, -23.06, 67.24, 99.95, 9.41E-02,  0.0,0.000,  0.00, 1.81, 3.220,
  0.00, 0.00,    0.00,    0.00,     1.94, 241.55E+07, 342.19E+01, 196.61E+01, 0.00E+00, 0.00E+00,
434., ast/iron, 2.98E+09, 0.00E+00, 12.75, 2.42E+21, 4.37E-10,  ,If , 59.94,  1.34,  12.08, 136.66,  10.73, -28.17, 66.39, 99.94, 5.19E-02,  0.0,0.000,  0.00, 2.53, 2.353,
  0.00, 0.00,    0.00,    0.00,     1.63, 401.60E+07, 906.83E+00, 188.84E+00, 858.96E-01, 0.00E+00,
472., ast/ach , 3.25E+10, 0.00E+00, 23.43, 8.91E+22, 6.35E-12,  ,If , 71.10, 10.25,  23.24, 115.98,   4.95, -24.14, 72.56, 99.94, 2.09E+00,  0.0,0.000,  0.00, 0.00, 1.657,
  0.00, 0.00,    2.52,    0.00,     9.58, 278.12E+07,          0.00E+00, 130.90E+01, 0.00E+00, 0.00E+00,
532., ast/iron, 2.79E+11, 0.00E+00, 26.77, 9.98E+23, 1.91E-07,  ,If , 21.20, 11.84,  26.60, 316.08,  11.82, -29.48, 18.94, 99.96, 2.36E+01,  0.0,0.000,  0.00, 17.28, 1.665,
  0.00, 0.00,    0.00,    0.00,    39.28, 234.81E+08, 866.29E+01, 223.90E+02, 0.00E+00, 0.00E+00,
625., ast/iron, 8.69E+09, 0.00E+00, 15.05, 9.84E+21, 5.95E-09,  ,If , 39.87,  1.75,  14.30, 185.45,  12.35, -28.98, 41.69, 99.93, 2.12E-01,  0.0,0.000,  0.00, 4.38, 1.245,
  0.00, 0.00,    0.00,    0.00,     3.61, 961.60E+07, 211.69E+00, 718.24E-01, 0.00E+00, 0.00E+00,
630., ast/ach , 1.17E+10, 0.00E+00, 12.08, 8.54E+21, 2.29E-12,  ,If , 42.13,  3.48,  11.40, 173.83,  14.42, -22.48, 45.02, 99.95, 1.82E-01,  0.0,0.000,  0.00, 0.55, 2.013,
  0.00, 0.00,    0.00,    0.00,     2.55, 210.02E+07, 197.94E-01, 210.59E+00, 0.00E+00, 0.00E+00,
667., ast/Och , 1.11E+11, 0.00E+00, 16.72, 1.55E+23, 1.09E-09,  ,If , 64.88, 12.38,  16.68, 118.89,   7.11, -24.59, 65.26, 99.99, 3.69E+00,  0.0,0.000,  0.00, 0.00, 1.848,
  0.00, 0.00,    0.00,    0.00,    13.33, 336.54E+07,          0.00E+00, 393.52E+01, 0.00E+00, 0.00E+00,
750., ast/Och , 1.22E+11, 0.00E+00, 18.70, 2.13E+23, 1.19E-09,  ,If , 61.72, 15.40,  18.68, 118.90,   6.35, -24.52, 61.73, 99.99, 5.07E+00,  0.0,0.000,  0.00, 0.00, 0.827,
  0.00, 0.00,    0.00,    0.00,    15.19, 298.41E+07,          0.00E+00, 486.83E+00, 0.00E+00, 0.00E+00,
842., ast/pall, 2.47E+10, 0.00E+00, 13.68, 2.31E+22, 1.45E-11,  ,If , 13.01,  2.95,  11.67, 591.14,  43.56, -26.79, 11.40, 99.87, 4.01E-01,  0.0,0.000,  0.00, 4.61, 0.398,
  0.00, 0.00,    0.00,    0.00,     4.82, 644.34E+07, 333.82E-01, 182.22E-01, 0.00E+00, 0.00E+00,
935., ast/ach , 1.38E+10, 0.00E+00, 28.29, 5.50E+22, 2.60E-12,  ,If , 64.05, 13.22,  28.09, 118.88,   4.20, -23.89, 64.62, 99.90, 1.30E+00,  0.0,0.000,  0.00, 0.00, 1.067,
  0.00, 0.00,    0.00,    0.00,     1.72, 103.47E+07,          0.00E+00, 109.08E-01, 0.00E+00, 0.00E+00,
984., ast/Och , 1.21E+10, 0.00E+00, 23.83, 3.42E+22, 1.77E-11,  ,If , 79.09,  6.01,  23.52, 115.96,   4.87, -28.96, 84.81, 99.92, 7.97E-01,  0.0,0.000,  0.00, 0.00, 0.813,
  0.00, 0.00,    0.00,    0.00,     6.06, 307.84E+07,          0.00E+00, 751.05E-01, 0.00E+00, 0.00E+00,
Big-3 Masses:            138., com/per  , 3.18E+12, 23.47, 8.76E+24, 6.36E+05, 1.24E-08,
                         543., ast/Och  , 7.50E+11, 31.82, 3.80E+24, 3.75E+05, 7.33E-09,
                         368., ast/iron , 4.31E+11, 17.98, 6.96E+23, 1.51E+05, 2.95E-09,
Big-3 Iridium Events:;   532., ast/iron , 2.79E+11, 26.75, 9.98E+23, 9.76E+06, 1.91E-07,
                         138., com/per  , 3.18E+12, 23.47, 8.76E+24, 6.36E+05, 1.24E-08,
                         367., ast/iron , 2.66E+10, 12.17, 1.97E+22, 3.99E+05, 7.80E-09,
Big-3 Energy Events:;    138., com/per  , 3.18E+12, 23.47, 8.76E+24, 6.36E+05, 1.24E-08,
                         638., com/long , 2.63E+11, 62.02, 5.06E+24, 2.63E+04, 5.15E-11,
                         543., ast/Och  , 7.50E+11, 31.82, 3.80E+24, 3.75E+05, 7.33E-09,
Fatalities for model 3008 (B,F,Ts,G): 4.99E+04, 1.15E+05, 0.00E+00, 8.59E+01,
```

Event Number	Type Offshore dist(km)	Mass (g) runup&t (km)	Velocity impact (km/s) fldPen (km)	Energy (erg) R(1psi) (km)	Ir M/A (ug/cm2) FluExpl erg/cm2	B? Raff bFatal	I? B fsFatal	Elev F$ (deg) gFatal	Zmin (km)	Vfinal (km/s)	L(burn) (km)	t(burn) (s)	Max mag	Elfinal (deg)	mass% left	Yield Mton	Xnox (ppm)	Blowoff FB	Crater r(km)	r(4psi) (km)	LS
21.,ast/Och	1.91E+09, 0.00,	0.00E+00, 0.00,	14.99,	2.15E+21, 1.62,37406.89,	1.87E-11, 199.23E+06,	,li 496.72E-01,	,	9.08, 0.00E+00,	0.00,	3.41,1191.82, 246.47E-01,		91.13,	-26.50,	11.89,	98.99,	2.71E-03,	0.0,0.000,	0.06,	0.00,	0.71,	2.196,
27.,ast/pall	7.45E+09, 0.00,	0.00E+00, 0.00,	24.96,	2.32E+22, 4.37E-12,		,if	81.28, 0.00E+00,	8.11,	24.60, 0.00E+00,	113.06,	4.53,	-27.62,	90.00,	99.88,	5.38E-01,	0.0,0.000,	0.00,	0.00,	0.00,	2.526,
35.,ast/Och	4.90E+10, 0.00,	0.00E+00, 0.00,	16.56,	6.72E+22, 2.14, 114.23E+07,	4.79E-10, 0.00E+00,	,if 483.56E+00,		,27.51, 0.00E+00,	14.00,	16.44, 0.00E+00,	234.88,	14.18,	-23.77,	26.39,	99.97,	1.58E+00,	0.0,0.000,	0.00,	0.00,	0.00,	1.040,
66.,ast/ach	2.03E+09, 0.00,	0.00E+00, 0.00,	11.88,	1.43E+21, 3.47, 112.56E+07,	3.97E-13, 0.00E+00,	,if 415.44E-01,		,62.71, 0.00E+00,	0.79,	10.56, 0.00E+00,	133.07,	11.26,	-24.59,	90.00,	99.91,	2.71E-02,	0.0,0.000,	0.00,	0.00,	2.24,	2.365,
94.,ast/Och	7.42E+10, 0.00,	0.00E+00, 0.00,	13.16,	6.43E+22, 5.47, 222.72E+07,	7.26E-10, 0.00E+00,	,if 1.24, 602.92E+07, 733.42E+00, 112.47E+01,		,54.05, 0.00E+00,	9.76,	13.09, 0.00E+00,	136.28, 415.80E+00,	10.35,	-23.79,	54.51,	99.99,	1.52E+00,	0.0,0.000,	0.00,	0.00,	0.00,	1.591,
118.,ast/iron	1.64E+09, 0.00,	0.00E+00, 0.00,	12.81, 5.79,	1.34E+21, 1.27, 801.08E+07,	4.81E-10, 119.60E-01,	,if 190.31E-02,		,51.68,	0.69,	11.77,	151.59, 751.33E-02,	11.87,	-29.22,	58.63,	99.91,	2.71E-02,	0.0,0.000,	0.00,	0.00,	2.26,	0.572,
147.,com/per	1.80E+11, 0.00,	9.02E+10, 0.00,	25.62, 24.17,	5.92E+23, 252.79E+07,	7.05E-10, 0.00E+00,	,of 443.09E-02,		,41.81,	27.93,	25.62,	139.20,	5.43,	-23.78,	41.05,	99.96,	1.41E+01,	0.0,0.000,	0.00,	0.00, -2.316,
203.,ast/iron	1.78E+12, 3.84,	0.00E+00, 0.00,	11.77, 0.00,	1.23E+24, X607571.80,	1.74E-08, 455.86E+05, 214.22E-08,	,oi	,47.01, 0.00E+00,	9.90,	11.76, 262.69E+02,	150.80, 0.00E+00,	12.79,	-25.68,	47.05,	100.00,	2.94E+01,	0.0,0.000,	1.01,	22.03,	1.308,
1683.81, 2891.07,	3.03E+09, 0.00,	0.00E+00, 0.00,	12.70,	2.44E+21, 3.47, 112.70,	2.07E-12, 0.00E+00,	,if 48.18,		,0.72,	11.82,	159.81,	12.62,	-29.46,	53.06,	99.93,	5.05E-02,	0.0,0.000,	0.00,	2.76,	2.182,
219.,ast/iron	3.70E+11, 0.00,	0.00E+00, 7.00,	21.13, 0.00,	8.26E+23, 3.62E-09, 1.79, 134.60E+08, 725.26E+00,		,if 152.74E+00,		,74.25, 0.00E+00,	16.72,	21.13, 447.05E+00,	107.30,	5.07,	-25.50,	74.50,	99.99,	1.97E+01,	0.0,0.000,	0.00,	2.06,	1.149,
238.,ast/Och	2.08E+11, 0.00,	0.00E+00, 0.00,	20.93, 34.81,	4.55E+23, 984.99E+07,	2.03E-09, 375.43E-01,	,if 536.13E+01,		,52.09, 0.00E+00,	17.41,	20.92, 0.00E+00,	130.50,	6.23,	-25.02,	51.68,	99.99,	1.09E+01,	0.0,0.000,	0.00,	0.00,	1.894,
243.,ast/Och	1.70E+09, 0.00,	0.00E+00, 0.00,	15.04, 24.38,	1.92E+21, 500.36E+07,	1.16E-10, 0.00E+00,	,if 146.31E+02,		,47.10, 0.00E+00,	1.16,	13.84, 0.00E+00,	162.16,	10.82,	-28.88,	52.28,	99.87,	3.88E-02,	0.0,0.000,	0.00,	2.36,	1.787,
384.,ast/iron	3.04E+11, 0.00,	0.00E+00, 0.00,	3.93, 0.00,	1.41, 402.42E+07,	213.99E+00,	,if 383.11E-01,		,16.88, 0.00E+00,	32.50,	25.49, 0.00E+00,	324.80, 567.31E-01,	12.74,	-22.90,	14.40,	99.95,	2.36E+01,	0.0,0.000,	0.00,	0.00,	1.291,
393.,com/per	0.00,	0.00,	0.00,	1.52E+11, 33.11, 311.76E+07,	9.88E+23, 1.19E-09,	,if 673.56E+01,		, 0.00E+00,		0.00E+00,											

```
405.,ast/iron, 1.46E+11, 0.00E+00, 11.39, 9.46E+22, 9.98E-09,   ,1f ,67.20,   1.66, 11.28, 128.08, 11.23, -29.39, 69.62, 99.99, 2.22E+00, 0.0,0.000, 0.00, 9.14, 0.893,
      0.00,   0.00,   0.00,   0.00, 12.26, 112.25E+09, 410.56E+00, 369.16E+00,   0.00E+00,   0.00E+00,
447.,ast/Och , 6.72E+10, 0.00E+00, 13.03, 5.71E+22, 6.58E-10,   ,1f ,77.30,   9.47, 12.98, 113.09,  8.67, -23.73, 80.14, 99.99, 1.35E+00, 0.0,0.000, 0.00, 0.00, 1.956,
      0.00,   0.00,   0.00,   0.00,  6.97, 210.44E+07,   0.00E+00, 137.81E+01,   0.00E+00,   0.00E+00,
486.,ast/iron, 8.93E+10, 0.00E+00, 20.61, 1.90E+23, 2.62E-08,   ,1f ,75.41,   4.87, 20.46, 118.88,  5.77, -29.63, 77.56, 99.98, 4.46E+00, 0.0,0.000, 0.00, 12.05, 1.938,
      0.00,   0.00,   0.00,   0.00, 17.14, 262.32E+08, 791.15E+01, 799.75E+01,   0.00E+00,   0.00E+00,
607.,com/long, 9.46E+10, 7.10E+10, 29.92, 4.24E+23, 1.85E-10,   ,1f ,36.28,  31.83, 29.92, 150.80,  5.04, -23.29, 35.31, 99.94, 1.01E+01, 0.0,0.000, 0.00, 0.00, 1.817,
     13.97, 139.31E+07,   0.00E+00, 402.73E+01,   0.00E+00,   0.00E+00,
707.,ast/CMch, 1.58E+12, 1.58E+11, 25.82, 5.27E+24, 1.24E-08,   ,1f ,42.00,  23.77, 25.82, 145.00,  5.61, -25.92, 41.22, 99.99, 1.26E+02, 0.0,0.000, 0.00, 24.30, 1.590,
      0.00,   0.00,  91.29, 311.09E+08, 144.37E+02, 101.89E+03,   0.00E+00,   0.00E+00,
711.,ast/ach , 4.46E+09, 0.00E+00, 11.55, 2.97E+21, 8.73E-13,   ,1f ,53.44,   0.74, 10.42, 147.61, 12.84, -25.08, 63.30, 99.93, 5.78E-02, 0.0,0.000, 0.00, 2.88, 2.791,
      0.00,   0.00,   7.25,   1.92, 146.26E+08, 321.51E+01, 715.40E+00,   0.00E+00, 194.84E+01,
835.,com/long, 6.50E+10, 4.88E+10, 58.42, 1.11E+24, 1.27E-10,   ,1f ,35.78,  39.61, 58.41, 139.20,  2.38, -25.52, 34.80, 99.85, 2.65E+01, 0.0,0.000, 0.00, 0.00, 0.984,
      0.00,   0.00,   0.00,   0.00, 32.27, 235.44E+07,   0.00E+00, 315.11E+01,   0.00E+00,   0.00E+00,
892.,ast/meso, 5.53E+09, 0.00E+00, 11.76, 3.83E+21, 2.16E-10,   ,1f ,53.71,   1.67, 11.06, 146.36, 12.46, -27.03, 59.34, 99.95, 8.07E-02, 0.0,0.000, 0.00, 2.79, 3.233,
      0.00,   0.00,   0.00,   2.04, 406.40E+07, 839.25E+01, 223.87E+01,   0.00E+00,   0.00E+00,
932.,ast/Cich, 3.58E+11, 7.17E+10, 20.33, 7.41E+23, 2.10E-09,   ,1f ,43.11,  27.46, 20.33, 136.30,  6.70, -23.28, 42.48, 99.99, 1.77E+01, 0.0,0.000, 0.00, 0.00, 2.232,
     28.99, 327.41E+07,   0.00E+00, 450.71E+02,   0.00E+00,   0.00E+00,
Big-3 Masses:,          203.,ast/Och , 1.78E+12, 11.77, 1.23E+24, 8.89E+05, 1.74E-08, ,
                        707.,ast/CMch, 1.58E+12, 25.82, 5.27E+24, 6.33E+05, 1.24E-08, ,
                        234.,com/per , 6.37E+11, 24.40, 1.90E+23, 1.27E+05, 2.49E-09, ,
Big-3 Iridium Events:,  486.,ast/iron, 8.93E+10, 20.61, 1.90E+23, 1.34E+06, 2.62E-08, ,
                        203.,ast/Och , 1.78E+12, 11.77, 1.23E+24, 8.89E+05, 1.74E-08, ,
                        707.,ast/CMch, 1.58E+12, 25.82, 5.27E+24, 6.33E+05, 1.24E-08, ,
Big-3 Energy Events:,   707.,ast/CMch, 1.58E+12, 25.82, 5.27E+24, 6.33E+05, 1.24E-08, ,
                        234.,com/per , 6.37E+11, 24.40, 1.90E+23, 2.49E+05, ,
                        271.,ast/Cich, 2.99E+11, 29.73, 1.32E+24, 8.98E+04, 1.76E-09, ,

Fatalities for model 3009 (B,F,Is,G): 3.61E+04, 1.95E+05, 2.63E+04, 2.90E+03,
```

References

Ahrens, T.J., and Harris, A.W. (1992). Deflection and fragmentation of near-earth asteroids. *Nature (London)* **360,** 429–433.
Ahrens, T.J., and O'Keefe, J.D. (1983). Impact of an asteroid or comet in the ocean and extinction of terrestrial life. *Proc. Lunar Planet. Sci. Conf.* **13,** A799–A806.
American Institute of Aeronautics and Astronautics (1995). "Responding to the Potential Threat of a Near-Earth-Object Impact." AIAA, Washington, DC.
Anonymous (D. Brewster?) (1819). Account of meteoric stones, masses of iron, and showers of dust, red snow, and other substances, which have fallen from the heavens, from the earliest period down to 1819. *Edinburgh Philosophical Journal,* **1**(#2), 221–235.
Anonymous (1890). Our astronomical column. *Nature (London)* **43,** 89–90.
Bailey, M.E., Clube, S.V.M., and Steel, D.I. (1993). The unforeseen hazard—Multiple impacts. *Pap., 17th Erice Workshop Planetary Emergencies, 1993.*
Baker, R.M.L., Jr. (1958). Ephemeral natural satellites. *Science* **128,** 1211–1213.
Baldwin, B., and Sheaffer, Y. (1971). Ablation and breakup of large meteoroids during atmospheric entry. *J. Geophys. Res.* **76,** 4653–4668.

Borovicka, J., and Ceplecha, Z. (1992). Earth-grazing fireball of October 13, 1990. *Astron. Astrophys.* **257,** 323–328.

Bottke, W.F., Jr., Nolan, M.C., Greenberg, R., and Kolvoord, R.A. (1994). Collisional lifetimes and impact statistics of near-Earth asteroids. *In* "Hazards due to Comets and Asteroids" (T. Gehrels, ed.), pp. 337–357. University of Arizona Press, Tucson.

Brett, R. (1992). The Cretaceous-Tertiary extinction: A lethal mechanism involving anhydrite target rocks. *Geochim. Cosmochim. Acta* **56,** 3603–3606.

Brin, G.D. (1980). Three models of dust layers on cometary nuclei. *Astrophys. J.* **237,** 265–279.

Brin, G.D., and Mendis, D.A. (1979). Dust release and mantle development in comets. *Astrophys. J.* **229,** 402–408.

Buddhue, J.D. (1954). Meteorites that have struck people or places. *Meteoritics* **1,** 359.

Canavan, G.H., Solem, J.C., and Rather, J.D.G. (1994). Near-Earth object interception workshop. *In* "Hazards due to Comets and Asteroids" (T. Gehrels, ed.), pp. 93–124. University of Arizona Press, Tucson.

Cassidy, W.A., Villar, L.M., Bunch, T.E., Kohman, T.P., and Milton, D.J. (1965). Meteorites and craters of Campo del Cielo, Argentina. *Science* **149,** 1055–1064.

Ceplecha, Z. (1988). Earth's influx of different populations of sporadic meteoroids from photographic and television data. *Bull. Astron. Inst. Czech.* **39,** 221–236.

Ceplecha, Z. (1993). Meteoroid properties from photographic records of meteors and fireballs. *In* "Asteroids, Comets, Meteors" (V. Carusi, ed.), pp. 343–356. IAU, The Netherlands.

Ceplecha, Z., and McCrosky, R.E. (1976). Fireball end heights: A diagnostic for the structure of meteoric material. *JGR, J. Geophys. Res.* **81,** 6257–6275.

Ceplecha, Z., Spurny', P., Borovicka, J., and Keclikova, J. (1993). Atmospheric fragmentation of meteoroids. *Astron. Astrophys.* **279,** 615–626.

Chapman, C.R. (1993). Thinking about the impact hazard: Comparisons with natural disasters and accidents. *Pap., 17th Erice Workshop Planetary Emergencies, 1993.*

Chapman, C.R., and Morrison, D. (1994). Impacts on the Earth by asteroids and comets: Assessing the hazard. *Nature (London)* **367,** 33–40.

Chen, G., Tyburczy, J.A., and Ahrens, T.J. (1994). Shock-induced devolatilization of calcium sulfate and implications for K-T extinctions. *Earth Planet. Sci. Lett.* **128,** 615–628.

Chyba, C.F. (1991). Terrestrial mantle siderophiles and the lunar impact record. *Icarus* **92,** 217–233.

Chyba, C.F. (1993). Explosions of small spacewatch objects in the Earth's atmosphere. *Nature (London)* **363,** 701–703.

Chyba, C.F., Thomas, P.J., and Zahnle, K.J. (1993). The 1908 Tunguska explosion: Atmospheric disruption of a stony asteroid. *Nature (London)* **361,** 40–44.

References

Croft, S.K. (1982). A first-order estimate of shock heating and vaporization in oceanic impacts. *Spec. Pap.—Geol. Soc. Am.* **190**, 143–152.

Crutzen, P.J. (1987). Acid rain at the K/T boundary. *Nature (London)* **330**, 108–109.

Degewij, J., and Tedesco, E.F. (1982). Do comets evolve into asteroids? Evidence from physical studies. *In* "Comets" (L.L. Wilkening, ed.), pp. 665–695. University of Arizona Press, Tucson.

de Laubenfels, M.W. (1956). Dinosaur extinction: One more hypothesis. *J. Paleontol.* **30**, 207–212.

Dorman, J., Evans, S., Nakamura, Y., and Latham, G. (1978). On the time-varying properties of the lunar seismic meteoroid population. *Proc. Lunar Planet. Sci. Conf.* **9**, 3615–3626.

Drummond, J.D. (1990). Earth-approaching asteroid streams. *Icarus* **89**, 14–25.

Fanale, F.P., and Salvail, J.R. (1984). An idealized short-period comet model: Surface insolation. *Icarus* **60**, 476–511.

Fanale, F.P., and Salvail, J.R. (1987). The loss and depth of CO_2 ice in comet nuclei. *Icarus* **72**, 535–554.

Frank, L.A., and Sigwarth, J.B. (1993). Atmospheric holes and small comets. *Rev. Geophys.* **31**, 1–28.

Gallant, R.A. (1994). Journey to Tunguska. *Sky Telescope*, June, pp. 38–43.

Gault, D.E., and Sonett, C.P. (1982). Laboratory simulation of pelagic asteroidal impact: Atmospheric injection, Benthic topography, and the surface wave radiation field. *Spec. Pap.—Geol. Soc. Am.* **190**, 69–72.

Gersonde, R., Kyte, F.T., Bleil, U., Diekmann, B., Flores, J.A., Gohl, K., Grahl, G., Hagen, R., Kuhn, G., Sierro, F.J., Völker, D., Abelmann, A., and Bostwick, J.A. (1997). Geological record and reconstruction of the Late Pliocene impact of the Eltanin asteroid in the southern Ocean. *Nature (London)* **390**, 357–363.

Glancy, A.E. (1916). A luminous object suspected to be a comet. *Publ. Astron. Soc. Pac.* **28**, 179–182.

Glasstone, S., and Dolan, P.J. (1977). "The Effects of Nuclear Weapons." U.S. Govt. Printing Office, Washington, DC.

Graham, A.L., Bevan, A.W.R., and Hutchison, R. (1985). *Catalogue of Meteorites*, Univ. of Arizona Press, Tucson.

Grieve, R.A.F., and Shoemaker, E.M. (1994). The record of past impacts on Earth. *In* "Hazards due to Comets and Asteroids" (T. Gehrels, ed.), pp. 417–462. University of Arizona Press, Tucson.

Halliday, I., Blackwell, A.T., and Griffin, A.A. (1989). The typical meteorite event, based on photographic records of 44 fireballs. *Meteoritics* **24**, 65–72.

Harris, A.W., Canavan, G.H., Sagan, C., and Ostro, S.J. (1994). The deflection dilemma: Use versus misuse of technologies for avoiding interplanetary collision hazards. *In* "Hazards due to Comets and Asteroids" (T. Gehrels, ed.), pp. 1145–1155. University of Arizona Press, Tucson.

Hartmann, W.K., Tholen, D.J., and Cruikshank, D.P. (1987). The relationship of active comets, "extinct" comets, and dark asteroids. *Icarus* **69**, 33–50.

Hills, J.G., and Goda, M.P. (1993). The fragmentation of small asteroids in the atmosphere. *Astron. J.* **105**, 1114–1144.

Hills, J.G., Nemtchinov, I.V., Popov, S.P., and Teterev, A.V. (1994). Tsunami generated by small asteroid impacts. *In* "Hazards due to Comets and Asteroids" (T. Gehrels, ed.), pp. 779–790. University of Arizona Press, Tucson.

Horanyi, M., Gombosi, T.I., Cravens, T.E., Koromezey, A., Kecskemety, K., Nagy, A.F., and Szego, K. (1984). The friable sponge model of a cometary nucleus. *Astrophys. J.* **278**, 449–455.

Houpis, H.L.F., Ip, W.H., and Mendis, D.A. (1985). The chemical differentiation of the cometary nucleus: The process and its consequences. *Astrophys. J.* **295**, 654–667.

Hudson, R.S., and Ostro, S.J. (1994). Shape of asteroid 4769 Castalia (1989 PB) from inversion of radar images. *Science* **263**, 940–943.

Hudson, R.S., and Ostro, S.J. (1995). Shape and non-principal axis spin state of asteroid 4179 Toutatis. *Science* **270**, 84–86.

Hughes, D.W. (1990). The mass distribution of comets and meteoroid streams and the shower/sporadic ratio in the incident visual meteoroid flux. *Mon. Not. R. Astron. Soc.* **245**, 198–203.

Jacchia, L.G. (1974). A meteorite that missed the Earth. *Sky Telescope* **48**, 4–6.

Jenniskens, P. (1994). The Mbale meteorite shower. *Meteoritics* **29**, 246–254.

Joule, J.P. (1848). On shooting stars. *Philos. Mag.* [3] **32**, 349–351.

Kisslinger, C. (1992). Sizing up the threat. *Nature (London)* **355**, 18–19.

Kring, D.A., Melosh, H.J., and Hunten, D.M. (1996). Impact-induced perturbations of atmospheric sulfur. *Earth Planet. Sci. Lett.* **140**, 201–212.

Krinov, E.E. (1966). "Giant Meteorites" (J. Romankiewitz, transl.). Pergamon, Oxford.

Lewis, J.S. (1996). "Rain of Iron and Ice." Addison-Wesley, Reading, MA.

Lewis, J.S. (1997). "Rain of Iron and Ice," rev. ed. Addison-Wesley, Reading, MA.

Lewis, J.S., Watkins, G.H., Hartman, H., and Prinn, R.G. (1982). Chemical consequences of major impact events on Earth. *Spec. Pap.—Geol. Soc. Am.* **190**, 215–221.

Love, S.G., and Brownlee, D.E. (1993). A direct measurement of the terrestrial mass accretion rate of cosmic dust. *Science* **262**, 550–553.

MacDougall, J.D. (1988). Seawater strontium isotopes, acid rain, and the Cretaceous-Tertiary boundary. *Science* **239**, 485–487.

Marsden, B.G., and Steel, D.I. (1994). Warning times and impact probabilities for long-period comets. *In* "Hazards due to Comets and Asteroids" (T. Gehrels, ed.), pp. 221–239. University of Arizona Press, Tucson.

McCrosky, R.E., Posen, A., Schwartz, G., and Shao, C.-Y. (1971). Lost city meteorite—Its recovery and a comparison with other fireballs. *J. Geophys. Res.* **76**, 4090–4108.

References

McKay, C.P., Squyres, S.W., and Reynolds, R.T. (1986). Methods for computing comet core temperatures. *Icarus* **66**, 625–629.

McKinnon, W.B. (1982). Impact into the Earth's ocean floor: Preliminary experiments, a planetary model, and possibilities for detection. *Spec. Pap.—Geol. Soc. Am.* **190**, 129–142.

McLaren, D. (1970). Presidential address: Time, life, and boundaries. *J. Paleontol.* **44**, 801–815.

Mekler, Y., Prialnik, D., and Podolak, M. (1990). Evaporation from a porous cometary nucleus. *Astrophys. J.* **356**, 682–686.

Melosh, H.J. (1982). The mechanics of large meteoroid impacts in the Earth's oceans. *Spec. Pap.—Geol. Soc. Am.* **190**, 121–127.

Melosh, H.J. (1989). "Impact Cratering: A Geological Process." Clarendon Press, Oxford.

Melosh, H.J., Nemchinov, I.V., and Zetzer, Yu.I. (1994). Non-nuclear strategies for deflecting comets and asteroids. In "Hazards due to Comets and Asteroids" (T. Gehrels, ed.), pp. 1111–1132. University of Arizona Press, Tucson.

Melosh, H.J., and Nemchinov, I.V. (1993). Solar asteroid diversion. *Nature (London)* **366**, 21–22.

Melosh, H.J., and Vickery, A.M. (1989). Impact erosion of the primordial atmosphere of Mars. *Nature (London)* **338**, 487–489.

Mendis, D.A., and Brin, G.D. (1977). Monochromatic brightness variations of comets II. Core-mantle model. *Moon Planets* **17**, 359–372.

Morrison, D., and Teller, E. (1994). The impact hazard: Issues for the future. The impact hazard. In "Hazards due to Comets and Asteroids" (T. Gehrels, ed.), pp. 1135–1144. University of Arizona Press, Tucson.

Morrison, D., Chapman, C.R., and Slovic, P. (1994). The impact hazard. In "Hazards due to Comets and Asteroids" (T. Gehrels, ed.), pp. 59–91. University of Arizona Press, Tucson.

Nagasawa, N., and Miura, K. (1987). Aerial path determination of a great fireball from sonic boom records on seismographs. *Bull. Earthquake Res. Inst., Univ. Tokyo* **62**, 579–588.

Nemtchinov, I.V., and Svetsov, V.V. (1991). Global consequences of radiation impulse caused by comet impact. *Adv. Space Res.* **11**(6), 95–97.

Nemtchinov, I.V., Svetsov, V.V., Kosarev, I.B., Golub', A.P., Popova, O.P., Shuvalov, V.V., Spalding, R.E., Jacobs, C., and Tagliaferri, E. (1997). Assessment of kinetic energy of meteorids detected by satellite-based light sensors. *Icarus* **130**, 259–274.

Neukum, G., and Ivanov, B.A. (1994). Crater size distributions and impact probabilities on earth from lunar, terrestrial-planet, and asteroid cratering data. In "Hazards due to Comets and Asteroids" (T. Gehrels, ed.), pp.359–416. University of Arizona Press, Tucson.

Öpik, E.J. (1958a). "Physics of Meteor Flight in the Atmosphere." Interscience, New York.

Öpik, E.J. (1958b). On the catastrophic effects of collisions with celestial bodies. *Ir. Astron. J.* **5**, 34–36.
Ostro, S.J., Campbell, D.B., Chandler, J.F., Hine, A.A., Hudson, R.S., Rosema, K.D., Winkler, R., and Yeomans, D.K. (1991). Asteroid 1986 DA: Rare evidence for a metallic composition. *Science* **252**, 1399–1404.
Ostro, S.J., Hudson, R.S., Jurgens, R.F., Rosema, K.D., Campbell, D.B., Yeomans, D.K., Chandler, J.F., Giorgini, J.D., Winkler, R., Rose, R., Howard, S.D., Slade, M.A., Perillat, P., and Shapiro, I.I. (1995). Radar images of asteroid 4179 Toutatis. *Science* **270**, 80–83.
Park, C. (1978). Nitric oxide production by Tunguska meteor. *Acta Astronau.* **5**, 523–542.
Passey, Q.R., and Melosh, H.J. (1980). Effects of atmospheric breakup on crater field formation. *Icarus* **42**, 211–233.
Perrine, C.D. (1916). A luminous object seen on May 4, 1916. *Publ. Astron. Soc. Pac.* **28**, 176–179.
Podolak, M., and Herman, G. (1985). Numerical simulations of comet nuclei. II. The effect of the dust mantle. *Icarus* **61**, 267–277.
Pope, K.O., Baines, K.H., Ocampo, A.C., and Ivanov, B.A. (1994). Impact winter and the Cretaceous/Tertiary extinctions: Results of a Chicxulub asteroid impact model. *Earth Planet. Sci. Lett.* **128**, 719–725.
Prialnik, D., and Bar-Nun, A. (1988). The formation of a permanent dust mantle and its effect on cometary Activity. *Icarus* **74**, 272–283.
Prialnik, D., and Mekler, Y. (1991). The formation of an ice crust below the dust mantle of a cometary nucleus. *Astrophys. J.* **366**, 318–323.
Prinn, R.G., and Fegley, B., Jr. (1987). Bolide impacts, acid rain, and biospheric traumas at the Cretaceous-Tertiary boundary. *Earth Planet. Sci. Lett.* **83**, 1–15.
Rabinowitz, D.L. (1993). The size distribution of the Earth-approaching asteroids. *Astrophys. J.* **101**, 1518–1529.
Rabinowitz, D., Gehrels, T., Scotti, J.V., McMillan, R.S., Perry, M.L., Wisniewski, W., Larson, S.M., Howell, E.S., and Mueller, B.E.A. (1993). Evidence for a near-Earth asteroid belt. *Nature (London)* **363**, 704–706.
Rabinowitz, D.L., Bowell, E., Shoemaker, E., and Muinonen, K. (1994). The population of Earth-crossing asteroids. In "Hazards due to Comets and Asteroids" (T. Gehrels, ed.), pp. 285–312. University of Arizona Press, Tucson.
Rather, J.D.G., Rahe, J.H., and Canavan, G., eds. (1992). "Summary Report of the Near-Earth Object Interception Workshop." NASA, Washington, DC.
Rawcliffe, R.D. (1979). Satellite observations of meteors. *Astrophys. J.* **228**, 338–345.
Remo, J.L. (1994). Classifying and modeling NEO material properties and interactions. In "Hazards due to Comets and Asteroids" (T. Gehrels, ed.), pp. 551–596. University of Arizona Press, Tucson.

ReVelle, D.O. (1979). A quasi-simple ablation for large meteorite entry. *J. Atmos. Terr. Phys.* **41**, 453–473.
Rietmeijer, F.J.M. (1989). Extraterrestrial sulfur in the lower stratosphere contributed by chondritic interplanetary dust particles. *Meteoritics* **24**, 319–320.
Sagan, C., and Ostro, S.J. (1994). Dangers of asteroid deflection. *Nature (London)* **368**, 501.
Saidov, K.H., and Simek, M. (1989). Luminous efficiency coefficient from simultaneous meteor observations. *Bull. Astron. Inst. Czech.* **40**, 330–332.
Schenk, P., and Melosh, H.J. (1993). Crater chains on Callisto. *Nature (London)* **365**, 731–733.
Schultz, P.H., and Beatty, J.K. (1992). Teardrops on the pampas. *Sky Telescope*, April, pp. 387–392.
Schultz, P.H., and Gault, D.E. (1975). Seismic effects from major basin formation on the Moon and Mercury. *The Moon* **12**, 159–177.
Schultz, P.H., and Lianza, R.E. (1992). Recent grazing impacts on the Earth recorded in the Rio Cuarto Crater Field, Argentina. *Nature (London)* **355**, 234–237.
Scotti, J.V., Rabinowitz, D.L., and Marsden, B.G. (1991). Near miss of the Earth by a small asteroid. *Nature (London)* **354**, 287–289.
Sekanina, Z. (1982). The problem of split comets in review. *In* "Comets" (L.L. Wilkening, ed.), pp. 251–287. University of Arizona Press, Tucson.
Sekanina, Z. (1983). The Tunguska Event: No cometary signature in evidence. *Astron. J.* **88**, 1382–1414.
Shoemaker, E.H. (1962). Interpretation of lunar craters. *In* "Physics and Astronomy of the Moon" (Z. Kopal, ed.), pp. 341–343.
Shoemaker, E.M., Weissman, P.R., and Shoemaker, C.S. (1994). The flux of periodic comets near Earth. *In* "Hazards due to Comets and Asteroids" (T. Gehrels, ed.), pp. 313–336. University of Arizona Press, Tucson.
Sigurdsson, H., D'Hondt, S., and Carey, S. (1992). The impact of the Cretaceous/Tertiary bolide on evaporite terrane and generation of major sulfuric acid aerosol. *Earth Planet Sci. Lett.* **109**, 543–559.
Smette, A., and Hainault, O. (1992). A near miss? *ESO Messenger* **67**, 57.
Sonett, C.P., Pearce, S.J., and Gault, D.E. (1991). The oceanic impact of large objects. *Adv. Space Res.* **11**(6), 77–86.
Spratt, C.E. (1991). Possible hazards of meteorite falls. *J. R. Astron. Soc. Can.* **85**, 263–280.
Spratt, C.E., and Stephens, S. (1992). Meteorites that have struck home. *Mercury*, March/April, pp. 50–56.
Steel, D.I., Asher, D.J., and Clube, S.V.M. (1991). The structure and evolution of the Taurid complex. *Mon. Not. R. Astron. Soc.* **251**, 632–648.
Streliz, R. (1979). Meteorite impact in the ocean. *Proc. Lunar Planet. Conf.* **10**, 2799–2813.
Svetsov, V.V. (1996). Total ablation of the debris from the 1908 Tunguska explosion. *Nature (London)* **383**, 697–699.

Swindel, G.W., Jr., and Jones, W.B. (1954). The Sylacauga, Talladega County, Alabama Aerolite: A recent meteoritic fall that injured a human being. *Meteoritics* **1**, 125–132.

Tagliaferri, E., Spalding, R., Jacobs, C., Worden, S.P., and Erlich, A. (1994). Detection of meteoroid impacts by optical sensors in Earth orbit. *In* "Hazards due to Comets and Asteroids" (T. Gehrels, ed.), pp. 199–220. University of Arizona Press, Tucson.

Toon, O.B., Zahnle, K., Morrison, D., Turco, R.P., and Covey, C. (1995). Environmental perturbations caused by the impacts of asteroids and comets. *Rev. Geophys.* **35**, 41–78.

Turco, R.P., Toon, O.B., Park, C., Whitten, R.C., Pollack, J.B., and Noerdlinger, P. (1981). Tunguska Meteor Fall of 1908: Effects on stratospheric ozone. *Science* **214**, 19–33.

Turco, R.P., Toon, O.B., Ackerman, T.P., Pollack, J.B., and Sagan, C. (1983). Nuclear winter: Global consequences of multiple nuclear explosions. *Science* **222**, 1283–1292.

Urey, H.C. (1973). Cometary collisions and geological periods. *Nature (London)* **242**, 32–33.

Vickery, A.M., and Melosh, H.J. (1990). Atmospheric erosion and impactor retention in large impacts, with application to mass extinctions. *Spe. Pap.—Geol. Soc. Am.* **247**, 289–300.

Wasson, J.T. (1974). "Meteorites: Classification and Properties." Springer-Verlag, Berlin.

Watkins, G.H., and Lewis, J.S. (1986). Evolution of the atmosphere of Mars as the result of asteroidal and cometary impacts. *In* "Workshop on the Evolution of the Martian Atmosphere" (M. Carr, P. James, C. Leovy and R. Pepin, eds.), LPI Tech. Rep. 86-07, p. 46. Lunar and Planetary Institute, Houston, TX.

Wetherill, G. (1989). Cratering of the terrestrial planets by Apollo objects. *Meteoritics* **24**, 15–22.

Wetherill, G.W., and ReVelle, D.O. (1982). Relationships between comets, large meteors, and meteorites. *In* "Comets" (L.L. Wilkening, ed.), pp. 297–322. University of Arizona Press, Tucson.

Wiegert, P.A., Innanen, K.A., and Mikhole, S. (1997). An asteroidal companion to Earth. *Nature (London)* **387**, 685–686.

Willoughby, A.J., McGuire, M.L., Borowski, S.K., and Howe, S.D. (1994). The role of nuclear thermal propulsion in mitigating Earth-threatening asteroids. *In* "Hazards due to Comets and Asteroids" (T. Gehrels, ed.), pp. 1073–1088. University of Arizona Press, Tucson.

Yau, K., Weissman, P., and Yeomans, D. (1994). Meteorite falls in China and some related human casualty events. *Meteoritics* **29**, 864–871.

Zahnle, K. (1990). Atmospheric chemistry by large impacts. *Spec. Pap.—Geol. Soc. Am.* **247**, 271–288.

Index

Ablation rate, impactors, 57, 62
Acceleration, calculation for
    impactors, 61
Apollo object, fatality modeling,
    128–129
Asteroid, *see also* Impactor;
    Near-Earth asteroid
    density, 51
    orbit deduction, 127–128
    ratio to number of comets, 46
Aten object, fatality modeling,
    128–129
Atmosphere
    erosion following impact, 73
    modeling, 60–61

Bible, historical records of
    impacts, 3
BLOWOFF, subroutine,
    85, 151
Brazil, historical records of
    impacts, 19

China, historical records of
    impacts, 4
Comet, *see also* Impactor;
    Short-period comet
    density, 50–51
    fatality of long-period comets,
        126–127, 130

Comet (*continued*)
  long-term predictions, 125–126
  ratio to number of asteroids, 46
  sizes, 45–46
  velocity distributions for impactors, 34–37
CRATER, subroutine, 85, 152
Crete, historical records of impacts, 3

DATAPRINT, subroutine, 87, 154
Deceleration, calculation for impactors, 61–62
Defense Support Program (DSP), impact flux data, 28–29
Deflection, potential impactors, 125, 134–136, 140
Density, impactors, 50–51
DSP, *see* Defense Support Program
Dust
  climatological effects, 70
  distribution modeling, 70

Eardrum, rupture, 77
Earthquake
  energy release, 32–34
  seismic wave from impact, 70–71
  triggering from impact, 81–82
ENERGY, subroutine, 84, 150–151
Energy dissipation, entering bodies, 62
Entry angle
  calculation, 60
  skipout, 60

Entry velocity
  chondrite impactors, 104–106
  distribution for impactors of various classes, 34–37
European Network, impact flux data, 28–29

Fatality
  causes, 77
  HAZARDS 5.5 modeling
    100×1-year runs
      causes of death, 97, 100, 144
      distribution by impactor type, 89–90, 92–93, 98–99, 143–144
      distribution by impactor yield, 95–97
      frequency distribution, 89–90, 100–101
    1000×1-year runs
      distribution by impactor type, 102–104, 106, 109, 112, 114–115
      distribution by impactor yield, 106–110, 114
      frequency distribution, 102, 109, 111, 113–116
    dependence on duration of exposure to bombarding flux, 120–121
    overview, 86
    small objects, 128–129
  historical records of impacts, 1, 6, 19, 22
  long-period comets, 126–127, 130
  time-averaged risk, 25

Index

Finds, meteorite composition compared to falls, 43–45
Flux, *see* Impact flux
FORMPRINT, subroutine, 83, 148–149
FRAGMENT, subroutine, 85, 151–152
Fragmentation
  dispersal velocity of swarm, 69
  modeling, 63, 85
France, historical records of impacts, 3, 6, 91

Global killer, characterization and tracking, 130–132, 145
Government, loss of command structure effects, 81

HAZARD, subroutine, 85–86, 152–153
Hazardous materials, release on impact, 81
HAZARDS 5.5
  100 × 1-year runs
    altitude distribution by impactor type, 90, 94
    fatalities
      causes of death, 97, 100, 144
      distribution by impactor type, 89–90, 92–93, 98–99, 143–144
      distribution by impactor yield, 95–97
      frequency distribution, 89–90, 100–101
      historical record correlation, 91

  1000 × 1-year runs
    chondrite impactors, survival to ground, 104–106
    fatalities
      distribution by impactor type, 102–104, 106, 109, 112, 114–115
      distribution by impactor yield, 106–110, 114
      frequency distribution, 102, 109, 111, 113–116
      historical record correlation, 115
    impactor types and yields, 102
  code listing, 155–168
  distribution of events, 87–88
  long run outcomes, 116, 118–121, 144–145
  main program organization, 148
  output sample and headings, 169–185
  subroutines
    BLOWOFF, 85, 151
    CRATER, 85, 152
    DATAPRINT, 87, 154
    ENERGY, 84, 150–151
    FORMPRINT, 83, 148–149
    FRAGMENT, 85, 151–152
    HAZARD, 85–86, 152–153
    INITIALIZE, 83, 148–149
    KIND, 84, 150
    MASS, 83, 149
    SORT, 87, 153
    SUMMARY, 148, 154
  target planet selection, 147
  time spans, 83, 87
History, comet and asteroid impacts
  Bible, 3

History (*continued*)
  Brazil, 19
  casualties, 1, 6, 19, 22
  China, 4
  close calls, 19–21
  comparison to HAZARD modeling, 91, 115
  Crete, 3
  France, 3, 6, 91
  Italy, 4
  language problems in record interpretation, 2
  Mexico, 48–49
  ships at sea, 4–5
  Siberia, 6, 19
  table of historical accounts, 7–18
  United States, 22
Human population, *see* Population, human

Impact flux
  astronomical observation data, 28
  classification of impactors, 34, 39
  cratering data, 27
  Defense Support Program data, 28–29
  earthquake energy release, 32–34
  European Network data, 28–29
  flux versus yield curve modeling, 30, 32
  mass calculation for impactor, 30–32
  orbit deduction, 30, 124
  Prairie Network data, 28–29
  velocity distributions for impactors, 34–37
Impactor, *see also* Near-Earth asteroid; Short-period comet
  classification, 34, 39
  composition
    dustball bodies, 45
    finds versus observed falls, 43–45
    iridium, 47
    iron, 46
    metals, 49–50
    nickel, 54–55, 57
    strength effects, 52, 54–55
    sulfur, 48–49
    water, 50
  physical properties
    ablation rate, 57
    density, 50–51
    luminous efficiency, *see* Luminous efficiency
    overview by impactor type, 56
    strength, 51–52, 54–55, 57
  spectral characterization, 40–42
  structural diversity, 141
  velocity distributions, 34–37, 84
INITIALIZE, subroutine, 83, 148–149
Interception, destruction of potential impactors, 135, 140
Iridium, impactor composition, 47
Iron
  impactor composition, 46
  strength effects, 52, 54
Italy, historical records of impacts, 4

Joule, contributions to entry physics, 23

Index

KIND, subroutine, 84, 150

Local threat, characterization and tracking, 133–134
Long-period comet, *see* Comet
Luminous efficiency
   definition, 57
   flux calculation, 63
   iron dependence, 47–48
   metal effects, 49–50
   size dependence, 29

Mass
   calculation for impactor, 30–32
   comets, 45–46
MASS, subroutine, 83, 149
Mexico, historical records of impacts, 48–49

NEA, *see* Near-Earth asteroid
Near-Earth asteroid (NEA)
   1979 VA, 43
   1996 JA1, 21
   1997 XF11, 21
   dark asteroids, 42
   efficiency of discovery, 21–22, 25, 42, 131
   history, 19–21
   orbit deduction, 30
   radar characterization, 124–125
   size detection and modeling accuracy, 129–130
   spectral characterization, 40–42, 141
   warning time for impact, 22, 140

Nickel
   impactor composition, 54–55, 57
   strength effects, 54
Nitric oxide, generation by impactor and calculations, 64–65
Nitrous acid, impactor generation, 65

Öpik, contributions to entry physics, 23
Orbit
   deduction, 30, 124–128
   deflection of potential impactors, 125, 134–136

Periodic comet, *see* Short-period comet
Population, human
   density modeling, 76
   hazards of impact, overview, 77
   seacoast urbanization, 76, 132–133
Prairie Network, impact flux data, 28–29
Property damage, risk assessment and costs, 25, 79–80

Radar, characterization of potential impactors, 124–125
Radiant efficiency, *see* Luminous efficiency
Regional hazard, characterization and tracking, 132–133
Risk assessment, *see also* HAZARDS 5.5; Fatality
   complications, 25

Risk assessment (*continued*)
  global killers, 130–132
  local threats, 133–134
  property damage, 25, 79–80
  regional hazards, 132–133
  statistical averaging, 24

Search and tracking
  cost and benefits, 145–146
  efficiency of discovery, 21–22, 25, 42, 131
  global killers, 130–132
  local threats, 133–134
  regional hazards, 132–133
  size detection and modeling accuracy, 129–130
  Spacewatch sensitivity of object size detection, 28, 40
Seismic wave, *see* Earthquake
Ships at sea, historical records of impacts, 4–5
Shock wave
  altitude dependence, 66–67
  modeling, 138
Short-period comet
  composition, 42–43
  entry velocity, 34–35
  orbits, 42
Siberia, historical records of impacts, 6, 19
Skipout, trajectories, 60
Sonic boom, trajectory calculation of entering bodies, 63
SORT, subroutine, 87, 153
Spacecraft
  costs for object characterization, 142

  deflection of potential impactors, 135–136
Spacewatch, sensitivity of object size detection, 28, 40
Starvation, impact outcome, 82
Strength, impactors, 51–52, 54–55, 57
Sulfur, impactor composition, 48–49
Sulfur dioxide, release on impact and effects, 48–49, 65–66, 138–139
SUMMARY, subroutine, 148, 154

Tidal wave
  energy decay, 78
  impactor size dependence, 78–79
  modeling, overview, 71–73, 86, 137–138
  pulsating cycles, 77–78
  velocity, 78
Tsunami, *see* Tidal wave

United States, historical records of impacts, 22

Velocity distribution, impactors of various classes, 34–37, 84

Warning time, potential impact, 22, 140
Water, *see also* Tidal wave
  impactor composition, 50
  vapor effects following impact, 71
Weibull strength law, 51–52, 69